Engineering Measurements
Methods and Intrinsic Errors

Engineering Measurements
Methods and Intrinsic Errors

T A Polak and C Pande

Professional Engineering Publishing Limited
London and Bury St Edmunds, UK

First Published 1999

ISBN 1 86058 236 2

A CIP catalogue record for this book is available from the British Library.

Printed and bound in Great Britain by St Edmundsbury Press Limited,
Suffolk, UK

RELATED TITLES OF INTEREST

Title	**Editor/Author**	**ISBN**
An Introductory Guide to Flow Measurement	Roger Baker	0 85298 670 X
IMechE Engineers' Data Book	Clifford Matthews	1 86058 175 7
Journal of Strain Analysis for Engineering Design		ISSN 0309/3247
Vehicle Noise and Vibration	IMechE Conference Transactions	1 86058 145 5

CONTENTS

ACKNOWLEDGEMENTS

The authors wish to thank the following:

Peter Polak for his numerous contributions, his valued advice, and for initiating the whole project.

Plint and Partners Ltd, friends, and colleagues for supplying a wide variety of material.

David Spriggs of KLA-Tencor for the case study 'Position sensing of photo masks using a semiconductor sensor' and the example 'Computer simulation of an optical system'.

Graham Whyte for his case study on 'Borehole probes'.

Nick Osborne for the case study on a 'Charpy impact tester'.

Pete Lancaster of Keithley Instruments and Peter Price for their contributions on computer control and data acquisition.

David Hammond for his case study 'Fatigue testing of shot-peened specimens'.

Alison Schild of Hydrologic for her case study on 'Continuous measurement of flow in rivers by ultrasound'.

David Dowling of KLA-Tencor for his example 'Simulating release of fission gases in nuclear fuel rods'.

Tony Webb for contributions to the chapters on Electrical Measurements and Instrumentation and on Temperature Measurement.

Carol Bloxsome for her contributions on finite element analysis.

George Plint, Alex Alliston-Greiner, and Amir Davidov of Plint and Partners Ltd for reviewing the chapters on engine testing, friction and wear testing, and electrical measurements.

FOREWORD

This book sets out to cover most of the types of measurements regularly used by engineers. The sub-heading in the title '*Methods and Intrinsic Errors*' conveys the main thrust of the book, which is to alert the engineer to the wide variety of measurement methods and the ways in which they can go wrong. The errors which are covered include not only the inaccuracies of instruments themselves, but also the more serious systematic errors intrinsic to the method, which may be very large.

I have spent about half my working life in a laboratory context, starting before the days of strain-gauges, computers, fibre optics, and lasers. Very early on, I found a problem with weighing out samples of hygroscopic materials; these got heavier during weighing when using slow, non-electronic scales. Eventually a combination of speed and enclosure was found adequate for the purpose in hand. Later I came across many more problems and errors, and those which are relevant to current engineering practice are included in this book. The most recent one I spotted was a factor of two error shown in Fig. 2.1, which was published as part of an official report.

When I found my way into teaching I noted that my students, under ever-increasing lecture loads particularly in mathematics, did not seem to think about the pitfalls in laboratory and project work as deeply as I would have wished. Accordingly I wrote a short student-oriented book on this topic, which has long been out of print. I am very pleased that the authors have used some of this early material.

They have produced a work which is relevant to all practising engineers, almost amounting to an encyclopedia of methods and errors.

Peter Polak
Retired Senior Lecturer
Mechanical Engineering
University of Sheffield, UK
July 1999

AUTHORS' PREFACE

Just as the responsibility of the machine designer is to imagine all the possible ways in which his machine may fail, and to design accordingly, so the task of anyone taking measurements should include finding out how the measurements may go wrong, thereby avoiding substantial errors. There are also occasions where even small errors can be important, for example in a temperature / thermodynamics / calorimetry / gas analysis context, where measurements may form the basis of large contractual payments. If this book helps engineers to avoid some serious errors in measurements, then it will have achieved its purpose.

The book is aimed at a broad spectrum of engineers: those working in manufacturing and process industries, as well as those involved in research and development. It will also be useful to engineering students as an introduction to the wide range of measurement techniques currently available. The contents are almost entirely from the experience of the authors and various contributors – all the examples are real (though sometimes rendered anonymous to avoid embarrassment), and many of the errors described are large, ranging from 10% or so to a factor of 2 or more.

Chapter 2 – The Human Element – can be regarded as light general reading, although the types of errors dealt with are potentially serious and very real. Subsequent chapters deal with various common classes of measurement, with later chapters 9 (Surface Profile, Friction, and Wear Measurements), and 10 (Internal Combustion Engine Testing) being somewhat more specialized. The final chapter on the use of computers will inevitably be out of date in some respects by the time it is printed, but the general points and pitfalls will remain valid.

We have tried to be as comprehensive as possible in our coverage of measurement methods. Several new or emerging methods are included to make the reader aware of their existence, although for some of these there is as yet little knowledge on their potential pitfalls. We hope we have conveyed the need for an attitude of 'healthy scepticism' towards instrument readings. Armed with this, readers will be more likely to uncover problems in their own or others' work.

It must be emphasized that the errors described are substantially only the ones we and our contributors know about personally or are adequately documented in the references. It is in the nature of things that there will inevitably be more types which we have yet to discover.

We apologize in advance for any textual errors or omissions (no book on errors would be complete without them!), and would welcome feedback from readers regarding these or any suggestions for future improvements and enhancements.

Sandy Polak and Caroline Pande
July 1999

Chapter 1

Introduction

1.1 Purpose and scope

This book is intended for any engineer involved in measurements, which may be in manufacturing and process control, in development, troubleshooting, or research. Specialist knowledge is not a prerequisite, and the reader should gain insight into the equipment and methods commonly used, and develop a healthy scepticism about instrument readings.

The basic message of the book is that while an instrument may in itself be highly accurate, the way in which it is applied can render the results highly inaccurate. The aim is to help the engineer avoid, or at least minimize, these inaccuracies. Many of the commonly used measurement methods are described, together with their pitfalls and problems, interspersed with practical examples.

The authors have used illustrative practical examples from their own and their contributors' personal experience, which encompass design, development, research, manufacturing, and troubleshooting, in a variety of industries.

Nearly all measurements involve humans; even when their function has apparently been replaced by a computer or robot, the human is involved in the selection of the technique and in interpreting the results. Therefore, the first topic (Chapter 2) is the human element, ranging from simple mistakes (observer errors) to misinterpretations and confusions caused by badly presented data or instructions.

Chapters 3 to 6 deal with measurement of basic quantities including position, movement, force, pressure, temperature, and fluid flow. In each area the main techniques are described, and their limitations are explained and illustrated by examples. Physical problems with the sensors and transducers used are included, but the electrical side is not. Electrical measurements and instrumentation aspects of other measurements are covered in Chapter 7.

More specialized topics of material property measurements, surface profiles, friction and wear measurements, and engine testing are dealt with in Chapters 8 to 10.

A final chapter covers the use of computers, which are used for data acquisition and control of test equipment, and also as an analysis tool to extend, enhance, and sometimes replace, physical measurements.

Theoretical subjects such as the statistical treatment of errors are not included, as there are well covered elsewhere.

1.2 Errors – a discussion

Errors may be large, small, or insignificant. In many aspects of engineering, genuine high accuracy of measurement, by which we mean errors of less than 0.1%, is neither attainable nor necessary. (Obviously there are exceptions, such as in metrology where an error of even 0.1% would be unacceptably large.) However, many digital instruments and much reported data display four or more significant figures, giving a false impression of high accuracy.

This book mainly deals with large and medium errors. Large errors of 20% or more are usually fairly easy to detect by simple cross-checks, though finding the source of the error can still be tricky. Medium errors, in the range 5 to 10%, are much more difficult, as cross-checks may not reveal them. Errors of this magnitude can often only be eliminated by understanding the limitations of the method or equipment in use.

Small errors, of the order of 1% and less, are inherent in many of the measurement methods described in this book, and fortunately in many situations these small errors are not significant. Where an error of this magnitude is not acceptable, either the method must be changed to a more accurate one, or there must be regular *in situ* calibration against a known standard. A typical example is regular calibration of a load cell by applying a dead weight load *in situ*. Calibration off the machine is not sufficient as it eliminates effects of the mounting method, the machine environment, etc.

It is worth noting that small percentage errors can have a large effect on the final result, for example if the measurement is input into a calculation involving subtraction of two similar-sized quantities. The opposite can also occur, that a large error can have a small effect on the final result, depending on the mathematics involved. For these reasons it can be very instructive to carry out a sensitivity analysis on the effect of each variable, when a final quantity is calculated from one or several measurements.

Statistical techniques such as analysing the variance or standard deviation of a set of readings are often thought to give an indication of the accuracy of an instrument or measurement process. However, such techniques only deal with 'scatter' or random variations, and do not necessarily indicate anything about systematic errors. It is the systematic errors which are difficult to detect, and which are the main point of this book.

1.3 Terminology

Most engineers have acquired an intuitive understanding of various terms commonly used in measurements and instrumentation, such as 'meter', 'sensor', 'transducer' and there is no need for strict definitions of these here. However, there are various terms associated with errors, such as sensitivity, linearity, hysteresis, resolution, stability, accuracy, and several others, which may be understood less well. Definitions for some of these terms can be hard to find, so a short list of these is provided in the Appendix.

Chapter 2

The Human Element

This chapter is composed of miscellaneous factual examples, some taken from published results, others from the authors' solid experience in research, industrial, or consulting work. The errors in these examples have in general subsequently been corrected, but there are undoubtedly many more such errors in print which have never been detected or corrected.

The purpose of this chapter is to demonstrate that a sceptical attitude to presented information, and an enquiring mind when conducting experiments and observations, can be very valuable aids to avoiding gross errors.

It is noteworthy that whilst instrumentation errors often amount to only a few percent, errors of comprehension, reporting and computing, as noted below, may well exceed 100% and be difficult to recognize. This is the reason for presenting this apparently miscellaneous collection of examples at the beginning of the book rather than at the end.

2.1 Instrumentation

Humans are inherently involved in the process of measurements, even though computers and automated equipment are tending to reduce the human element. This trend almost certainly reduces the scope for human errors in the actual process of measuring and recording results. However, at the same time the possibilities for other errors may be increasing, as a consequence of the complexity of the equipment.

In the past, with mainly analogue instrumentation, an engineer would be expected to be able to operate a universal device, such as an oscilloscope, and to be aware of the obvious mistakes such as mis-selection of scales. Signal modifiers such as filters would be externally plugged in, and their presence would be obvious and unmistakable. Modern digital equivalent equipment may have a large number of useful features built-in, but the sheer number of options increases the possibility of accidentally using the wrong settings.

There is no simple answer to this, except perhaps a plea to equipment manufacturers to simplify operation, and perhaps to always have a baseline 'minimalist' mode with all the optional extras switched out. When the operator, in order to perform a simple operation, has to read a jargon-laden manual, the probability of error due to frustration and mental overload increases somewhat!

2.2 Input errors

One great input error occurred in world resource forecasting. Certain forecasts in 1971/2 showed an alarming divergence between population and resources. After further work it was found that a parameter had been typed in with the wrong order of magnitude. While correcting this and translating for a different computer, Boyle (**1**) reported further discrepancies and eventually published a much more comforting conclusion. Though not connected with the subject of this book, this example shows up the dangers in the highly demanding field of computing. The above work goes back many years, when the input mechanisms were still somewhat imperfect (punched cards and paper tape), capable of creating random errors when re-inserting or copying previously satisfactory programs. This should be much rarer with present magnetic storage methods; these days errors are likely to be made at the planning or keyboard stage.

There is no complete solution to this type of problem, but a general recommendation is to cross-check any complex computer model by means of a simplified hand calculation, or if not feasible, then by a simplified computer model written independently. The use of spread-sheets for modelling is helpful, in that input errors may be more readily visible.

2.3 Presentation errors

Presentation of information is relevant not only at the output stage of the measurement process, but also at the interpretation or analysis stage, where data from various sources may be combined or compared with the measurements taken. The following are some examples from the authors' experience; it is likely that the reader will have come across similar types of problems.

2.3.1 Misplaced captions

This error occurred in a report comparing various materials for strength and density, such as high-strength fibres, various steels, titanium, and aluminium alloys. The result looked somewhat like Fig. 2.1a. The caption suggested some improbably high strength values. The reason was an error in the layout commands which had mis-positioned the figures; they should have been as in Fig. 2.1b. Though the error would have been obvious to a specialist in modern materials, for a non-specialist the data appeared credible.

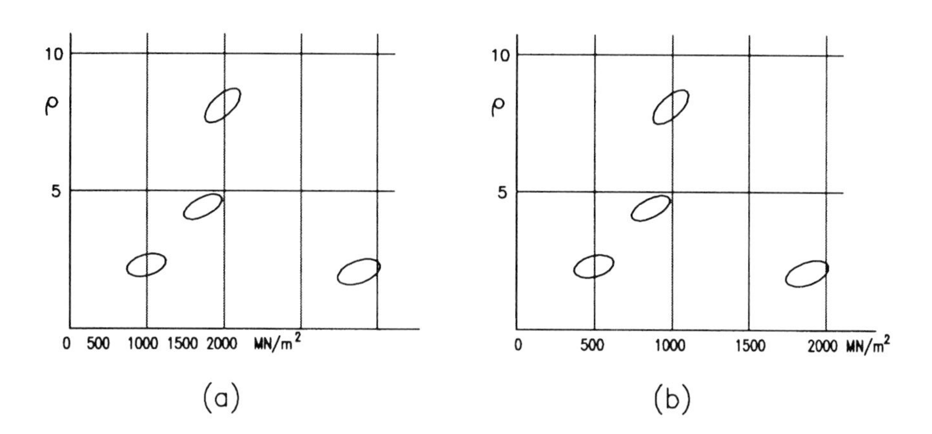

Fig. 2.1 Comparison of various materials for strength and density

2.3.2 Stresses and micro-strains

This describes an easily made error. A report based on strain gauge work referred to 'high stresses', whilst the actual data were in micro-strains (extensions in millionths) which is convenient and quite usual in this field. The misunderstanding is quite possible since Young's modulus in S.I. units for the material was around 2×10^5 N/mm^2, so that the micro-strain values would look quite credible if misread as stress values (in N/mm^2), having similar orders of magnitude but wrong by a factor of around 5.

2.3.3 Rotary speed

A similar problem occasionally occurs with rotary speed. Engineers are accustomed to revolutions per minute (rpm), and also to angular velocity in radians/second. Unfortunately, some design and selection calculation procedures use revolutions per second. If presented as Hz, there would be little danger, but when presented as r/s, there is a potential confusion with radians/second, with a discrepancy of a factor of 2π.

2.3.4 Acceleration and vibration measurements

In vibration measurements, acceleration may be presented in m/s^2, or in '*g*'s, i.e. multiples of 9.81 m/s^2. Further confusion is available in this field, as the figure may be amplitude (height from mean line to peak of waveform), peak-to-peak, or root mean square (RMS). The information confirming exactly what is being presented by a particular instrument is sometimes only found by careful perusal of the manual.

2.3.5 Rotary inertia

In presenting rotary inertia, British engineers and probably also North American engineers would expect to find this in terms of mass times radius of gyration squared, whereas the custom in Germany and some other European countries is to use mass times **diameter** of gyration squared. Whilst it is tempting to be prejudiced in favour of one's own custom, one should try to see a reason for the alternative view. The advantage of using the radius is that it gives the position of the equivalent mass directly which supports the physical viewing of the system concerned; also in conjunction with the angular velocity in radians/second, which is in keeping with physics practice, the radius gives immediate rise to the velocity, without a numerical factor. Perhaps the German custom arises linguistically since the word for radius is Halbmesser, which suggests that a halving is involved somehow. The error in mistaking one for the other is a factor of 2 squared, namely 4; the dimensions are the same so that there is no immediate signal to alert the reader. A quick cross-check using the mass and the approximate radius of the object (if known) is recommended.

2.3.6 Stress concentration data

One of the commonest stress-raisers is the junction between two shaft diameters, Fig. 2.2. Under load, the stress around the corner radius r is higher than in a plain shaft under similar loading, quoted as a stress concentration factor so that the designer calculates the stress for a plain shaft under the relevant condition and multiplies it by the stress concentration factor. Unfortunately some published factors use the stress in the larger shaft as basis while others use that in the smaller shaft. As a quick check, the factor for bending, based on the smaller shaft size, is roughly:

$$1 + 0.5\,(h/r)^{0.5}$$

where h is the step height $0.5(D - d)$.

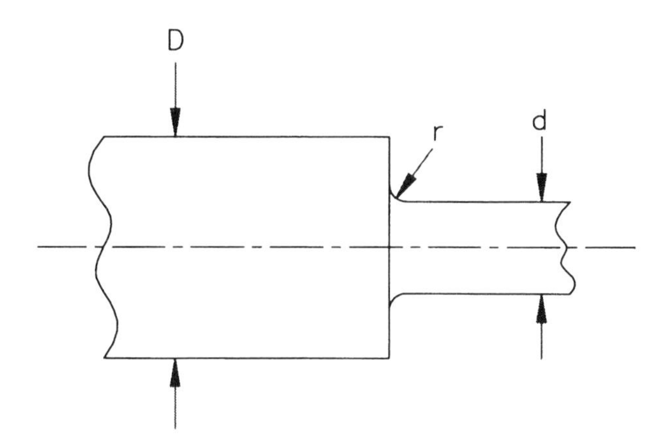

Fig. 2.2 Stress-raiser between two shaft diameters

2.3.7 Percentages

Reporting of large percentage changes is potentially ambiguous. For example, does 'an increase of 250%' mean 2.5 times the original, or 3.5 times? It depends entirely on how careful the writer has been with his use of English. If the source is a translation, the scope for ambiguity is even greater.

Expressions such as 'a 200% reduction' are meaningless and even if one guesses what the writer intended, one can easily be wrong. Even the apparently innocuous '70% reduction', while it should mean 0.3 times the original value, is sometimes used for 0.7 times the original value, or even (1/1.7) times, this last being the 'inverse' of a 70% increase!

The authors' view is that any 'increase' of over 100% is suspect, and any 'decrease' of 50% or more is also suspect. Wherever possible the actual figures should be checked.

2.4 Sources of data and information

To do useful and accurate development and testing work, the engineer must realize that the regular, long-established textbooks are just starting points, broad rather than deep, and in some ways become outdated over specific points. The next stage is provided by specialized texts from reputable publishers; these, however, are expensive and not always accessible. In many cases the most important sources are original papers in journals of the professional institutions; many are kept on reference in university libraries.

To find relevant papers, computer searching is widely accessible; other sources are index volumes of the appropriate journals covering groups of years and also the Engineering Index, a collection of annual volumes which embraces a wide range of engineering sources, periodicals as well as institution proceedings or society annals. The Engineering Index is noted for its excellent concise contents summaries, but it only includes periodicals, and not textbooks. As a further source one may try engineering handbooks which give short outlines and, what is more, appropriate references to further data. Another source of articles and references is provided by the encyclopaedias of science and technology.

Unfortunately, as the amount of technical information has grown, it has become more and more difficult to conduct a comprehensive search for data on any particular topic, particularly as valuable data in research theses are not adequately indexed anywhere accessible. Effective searches tend to be iterative, with the first round identifying particular authors or institutions, and the later rounds homing in on these sources. It is to be hoped that expanding use of the internet will make the process of searching more effective and easier.

Some warnings are called for in connection with technical articles and papers. Very early work did not have the benefit of instruments such as oscilloscopes or strain gauges. Another point, both good and bad, was a general acceptance that there was merit in conciseness and that the readers would not expect lengthy analyses or justifications. One result was that it was not easy to check the validity of the way conclusions were derived from the observations. In later work there has been the opposite tendency, avoiding the publishing of short papers which might be ignored as casual notes; therefore authors have adopted the remedy of presenting the same results in a number of different arrangements, which has some convenient aspects but gives the illusion of many more data than have been actually observed. Another trick of impression-making is to present computed values as points on a graph, whilst previously it was good practice to present calculated trends as lines on graphs (unless stated otherwise) and reserve the use of points for observed values.

2.5 Learning from the unexpected

While not strictly examples of errors, the following examples deal with chance observations and events during experimentation. To begin with, two examples from history show how some major advances started in this way; this will be followed by several minor items more in line with what engineers may come across in test work.

2.5.1 Newcomen's break

Savery and Newcomen may be considered the fathers of steam power. Savery's pump, which was still in industrial use this century, was a steam-powered water pump with no moving parts other than valves. The purpose was pumping water from mines. It consisted of a vessel with a steam inlet in the top provided with an inlet pipe and valve and also an outlet valve at the bottom feeding into a delivery pipe, to the mine surface, or to a similar vessel higher up (it is not certain whether this multi-stage system was actually used in deep mines but it is noteworthy that this method could achieve high lifts without high steam pressures). To start, steam was admitted to the vessel and then shut off. The condensing steam would cause the vessel to fill by suction (more correctly, by the pressure of the atmosphere); then steam at some pressure was admitted to blow the water out and upwards, the height being limited by whatever boiler pressure the designer considered safe.

Recognizing the dangers of combining high-pressure steam with large boilers, the Savery pump was wisely kept to small sizes. Moreover, the small size would favour fast condensation, thus giving frequent deliveries to make up for the smaller volume per cycle.

Newcomen saw a way of using the effect on a larger scale, for deeper mines without multi-staging, by using a vertical cylinder, a cross-beam and a series of long rods down to a pump at the base of the mine shaft. The vital step was to use the suction of condensing steam, in other words atmospheric pressure, to work the pump. The pump technology was already familiar but the making and sealing of large cylinders and pistons was at the limits of current techniques.

In each application the beam was ballasted according to the depth of the mine so that the rods would descend gently, aided by a slight steam pressure which helped to blow out any air (air would tend to slow down the condensation), but not enough to risk a boiler fracture. Many books on engineering thermodynamics show the layout.

The engines tended to be slow, the steam having to condense on the cylinder walls which were surrounded by a jacket supplied with cold water. The lucky break was a leak in the cylinder wall. This had the result, when the steam was shut off, of spraying cold water into the cylinder and vastly speeding up the condensing action. The effect was so rapid that the piston broke the chain attaching it to the beam, damaging the cylinder and even the top of the boiler below. (This occurred on a full-size prototype, not in service.) Thereafter Newcomen provided deliberate water injection, controlled by valves actuated from the beam, as also were the steam admission and cut-off arrangements. From then on, the Newcomen engines became a great success and were used in mines far and wide, in Britain, on the Continent and as far afield as New Jersey (in 1753). It is well known that this system was later superseded by the Watt system.

2.5.2 Bessemer's observation

Bessemer as a young man was helping to make steel in a reverberatory furnace; this involved turning over the pigs of cast iron so that the excess carbon could be burnt off by excess air in the flames. Using a wrought iron pole through a small access hole, he sometimes pierced a pig and noticed liquid iron running out; naturally enough, the oxidation proceeded from the surface inwards so that the pigs became more like steel on the surface, of higher melting point, while the high-carbon iron inside the shell remained liquid for longer. Realizing the significance of the oxidizing action eventually gave rise to the Bessemer converter, and after recognition of the problems posed by sulphur, phosphorus, etc., to successful bulk steel-making.

2.5.3 Apparent smoke reduction

Before the era of natural gas there were experiments going on to reduce smoke emission from domestic units using coal. A test unit for one such programme is shown in Fig. 2.3. The smoke density was measured by a light-beam and photo-cell. In one particular condition a virtually smoke-free state was observed. Fortunately, the laboratory was in temporary premises, with a glass roof and an individual flue from the apparatus concerned; by looking up one could see masses of smoke. The reason why the instruments showed no smoke was that the temperature in the observation section was so high that the tarry emissions were fully vaporized and transparent, whilst on entering the atmosphere they condensed into droplets. Incidentally, this example shows the importance of secondary air or lack of it.

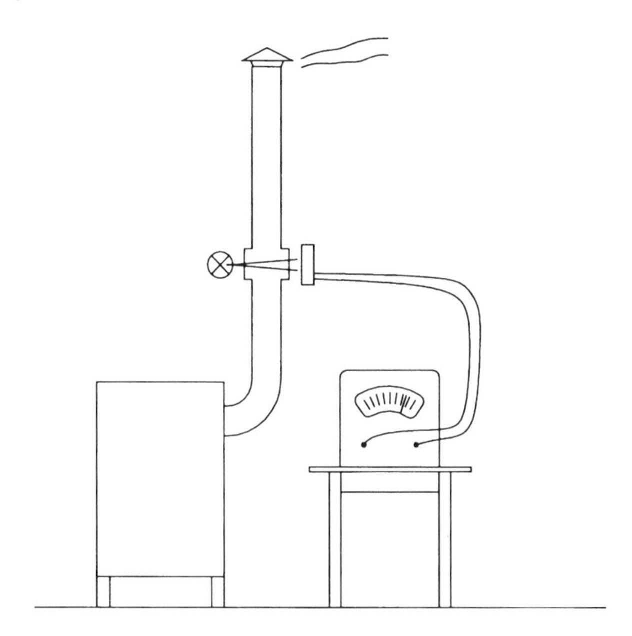

Fig. 2.3 Smoke emission test unit

2.5.4 Negative carbon dioxide?

In a combustion experiment with fairly wet fuel it was found that the exhaust gases showed unusually low carbon dioxide (CO_2) content, in one case even registering a negative value. The instrument (catharometer type) depended on the temperature reached by an electrically heated filament in a small chamber. On cross-checking by a different method it was found that when rather wet fuel was freshly added to the combustion chamber some hydrogen was generated, as in some gas producers. The thermal conductivity of CO_2 is lower than that of nitrogen or most other gases normally found in such circumstances, so the filament would usually become warmer in proportion to the amount of CO_2 but hydrogen has a high thermal conductivity, about eight times that of nitrogen, thus the presence of hydrogen grossly falsified the instrument reading. This problem was remedied by adding a copper oxide hot reducing section and a cooling section upstream of the CO_2 meter.

2.5.5 Calibration problem

On one occasion a propeller-type flowmeter, known as a current meter (for water) was being checked in a rotating calibration vessel, as illustrated in Fig. 2.4. To use the apparatus the chamber is set rotating at a fixed speed; after a time the water also reaches a constant speed; this can be confirmed by observing the height to which the water rises at the edge. Then the current meter is lowered into the water at the required depth of immersion and at the radius which gives the speed required. It was expected that the electrical count of propeller frequency would soon attain a steady reading.

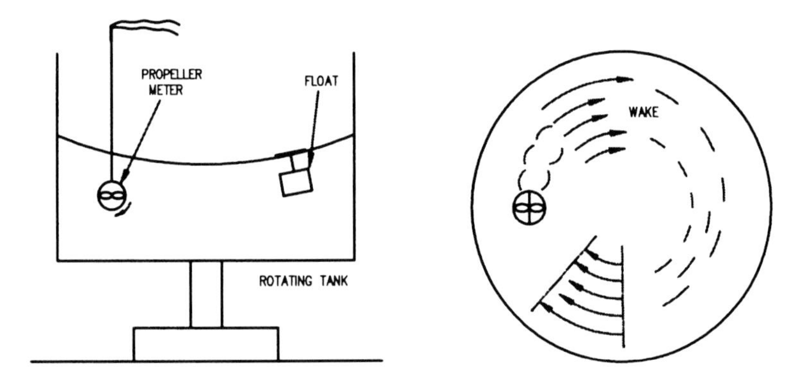

Fig. 2.4 Calibration problem with a propellor type flowmeter

However, the reading did not become steady but rose to a peak and then started diminishing. To investigate the problem, floats were made, with cork discs at the surface and extensions down to the level of measurement. One of these was dropped in just after the current meter, others at varying radii. It was found that the presence of the current meter significantly slowed down a region of the flow. In a case of this kind, the unattained condition when the meter is first put into the water can be estimated by taking a number of readings at known intervals, drawing a graph against time and extrapolating back to zero time. The method of extrapolating back from a series of readings is a valuable form of correcting for time-dependent errors and deserves to be more widely known. It is a useful technique to be applied when carrying out static calibration of certain transducer types which suffer from significant drift.

Chapter 3

Position, Speed, and Acceleration Measurement

This chapter is about measurement of position and movement, both linear and rotary. It does not specifically deal with measurement of physical dimensions (metrology), though some of the techniques described are also used in this field. A section on vibration measurement and interpretation is included.

3.1 Linear position

For manual measurement, traditional instruments such as micrometers, vernier scales, and dial indicators are all used, and give accuracy down to 5 or 10 μm. Electronic versions of these are now commonly used, with digital readouts and the capability to be linked into computers and dataloggers. The electronic versions usually give 2 to 4 times higher resolution than their purely mechanical equivalents.

The electronic micrometer or caliper is usually in the same form as its mechanical equivalent, but with a linear encoder unit incorporated. These typically use capacitance sensors, with multiple sender plates fed with alternating current (a.c.) voltages of different phases. The induced current on the receiver plate varies in phase as the sender and receiver plates move relative to one another. Early versions of these devices had a habit of giving errors if they were moved too rapidly, or if dirt contaminated the sensor surfaces. However, these problems have largely been eliminated with later designs.

Linear transducers vary enormously in terms of accuracy, range, and cost. Some of the principal types are described below.

Potentiometers are relatively low-cost, low-accuracy devices, with errors of the order of 1 or 2% typically. Higher accuracy versions are available, at higher cost. They are available in linear forms with operating ranges from a few mm to hundreds of mm.

Linear variable differential transformers (LVDTs) are commonly used and highly accurate. They consist of a core whose displacement gives rise to the signal or measurement by flux linking the energizing coil to the detecting coil, in proportion to the displacement.

The detecting coil is actually split, so that moving in a particular direction will increase the flux linkage with one, and decrease it with the other. The two halves are connected in opposition, as shown in Fig. 3.1, so that the output is the difference between the two signals (hence 'differential'). This doubles the sensitivity of the transducer. The output is then normalized by dividing by the sum of the two signals, and hence susceptibility to external influences such as temperature is decreased.

The excitation and signal processing may be either built into the instrument, or supplied externally. These are of linear rather than angular action, and achieve a fairly high degree of linearity, as well as resolution into the nanometer range. Often the resolution is limited only by the number of bits on the display used. They are available with ranges from less than 1 mm up to 100 mm or so.

Various types of proximity sensor exist for measuring small displacements, operating on inductive, capacitative, or Hall effect principles. These are only suitable for movements of at most a few mm, and typically 1 mm or less. One typical use is for monitoring the position of a shaft in a plain bearing, to indicate the thickness of the lubricating oil film. A mention of the use of a Hall effect sensor for measuring the lift of a diesel fuel injector needle is given in Chapter 10.

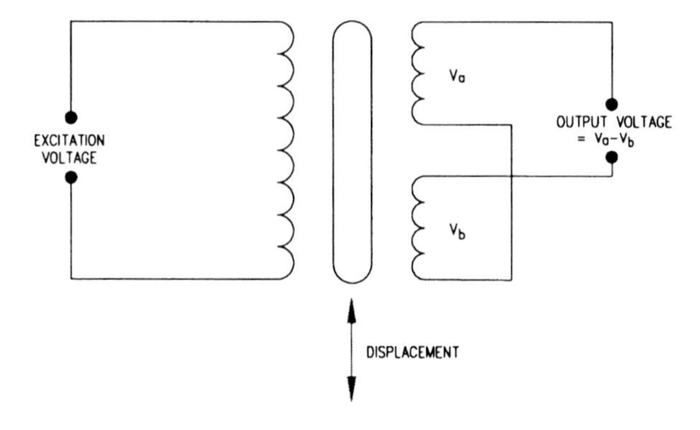

Fig. 3.1 The linear variable differential transformer (LVDT)

Another highly accurate non-contacting method is the silicon semiconductor position sensitive detector. An example of its use is given in the case study at the end of this chapter on 'Position sensing of photo masks using a semiconductor sensor'.

For larger distances, from hundreds of mm to tens or hundreds of metres, various forms of rangefinder exist. The oldest type is the optical rangefinder as used on older cameras, where two images are adjusted until they are superimposed, and the distance is read from a mechanical scale. Modern versions use infra-red, ultrasound, or radio waves (e.g. radio-altimeters, radar), detecting phase differences or time lags between the transmitted and reflected signals. All of these methods require a good target for reflecting the signal. It may be speculated that poor signal reflection from forested hillsides could upset radio-altimeters and contribute to some aircraft crashes.

None of these can be regarded as particularly precise, but they are often the only realistic options, and are sufficiently accurate for many purposes. A typical source of errors for the ultrasound method is temperature changes or gradients in the air, affecting the speed of transmission of the signal.

One example of the use of ultrasound distance measurement is for detecting the level of product in large tanks and silos. This generally works well for liquids, but for solids such as plastic granules a problem arises with 'coning', such that the silo empties from the middle outwards and the granules take up an inverted cone shape, so that the quantity of product may be significantly different from that predicted by the level measurement. The method has the merit that it can be easily retro-fitted to an existing installation, and is therefore often used, particularly for liquids.

For even larger distances, the Global Positioning System (GPS) can locate the absolute position of an object in the open air (i.e. an aircraft or a surface vehicle) to within 50 metres or so, by reference to a system of orbiting satellites. The military versions of this technology are believed to be considerably more accurate than this, but are not readily available for civilian use. The system gives vertical heights above sea level in addition to horizontal coordinates.

It is frequently convenient to convert linear distance or motion into rotary motion by means of a wheel, rack and pinion, or cable and drum, and then use a rotary measurement technique as described in Section 3.2.2. A typical example is a surveyor's wheel, a much more convenient device than the traditional chain.

3.2 Angular position and rotation

3.2.1 Level measurement and navigation

A simple form of angular position reference is the spirit level, often combined with an angular scale or protractor. The familiar builder's spirit level is not a particularly sensitive device, but reversing it can give reasonably accurate self-checking results. Levels used for surveying purposes (in theodolites and similar instruments) can be accurate to better than 1 minute of arc (0.3 milliradians). The spirit level works by gravity and therefore is subject to lunar and solar attraction. This has been calculated to cause a maximum error of less than 10^{-6} radians on approximately a daily cycle. This error is insignificant for most practical engineering purposes.

There is an electronic version of the spirit level, the fluid level inclinometer. The case study on 'Borehole probe' at the end of this chapter describes a particular application of this device. Various types of inclinometer exist, working either on a fluid level or a pendulum principle.

A magnetic compass is another familiar device, again typically of low accuracy of the order of a degree. Versions used for ships and aircraft can be more accurate, by a factor of 10 or so. Magnetic compasses are affected by large magnetic objects in the ground or even on the sea-bed, and also by magnetic materials and electromagnetic effects within the vehicle in which they are mounted. Quite small electric currents drawn by adjacent instruments can have a significant effect.

Many mechanical types of compass create large errors due to accelerations in an East–West direction. This is a consequence of the inclination to the horizontal of the earth's magnetic flux and the way in which the compass needle is suspended. This is why a small boat's or aircraft's compass may swing dramatically in rough conditions or while changing direction. Electronic 'flux-gate' type compasses and certain special mechanical types do not suffer from motion error.

Gyroscopes mounted in gimbals can provide an accurate angular position reference for only a short period. Friction in the gimbals and in the gyro bearings causes drift, and also the rotation of the earth requires compensating for. A gyro-compass is a combination of gyroscope and compass, with the compass compensating for the drift of the gyro, and the gyro stabilizing the short-term fluctuations of the compass.

3.2.2 Relative angles

The rotary potentiometer, as with the linear potentiometer, is an economical and commonly used angle transducer with an accuracy typically of the order of 1%. Higher accuracy can be achieved at higher cost. An interesting variant is the multiple turn potentiometer, often with 10 turns, which is a convenient way of measuring angular position varying by more than 360°, and which also gives higher accuracy of the order of 0.25%. The resistive track and wiper follow a helical path, but the drive shaft rotates normally.

Digital shaft encoders exist in various forms. Incremental encoders are essentially slotted discs with an optical detector which counts the slots and deduces the relative shaft position. These are used when the prime measurement is rotary speed, and are therefore described in a later section (3.4.2). The more certain method for angular position is the absolute shaft encoder which requires a lot more information than the incremental encoder.

In the absolute shaft encoder, the rotating element is a digitally encoded disc with several concentric tracks, and the output is a binary code. If it is in standard binary, then a single increment can mean that several of the digits have to change at once, so if one or more is slow to change then the output is momentarily entirely wrong. Some encoders use the Gray code, which avoids this problem. It has the advantage that it changes only one bit for each increment, so that if a bit is slow to change, then it just means the previous value will be present for a little longer, instead of giving a completely false reading. However, the need for Gray code has substantially disappeared with modern signal processing, which can detect and correct the type of error described above.

If it is only required to detect one or a few angular positions accurately, then a toothed or slotted steel wheel with a suitably positioned inductive sensor can be used. This can accurately detect the leading or trailing edge of each tooth, and this method is used for providing accurate timing signals for the spark on internal combustion (IC) engines, or for synchronizing various actions on high-speed machinery.

To ensure high accuracy with any of the above, the alignment must be good. The rotating component, if mounted directly on the shaft being measured, must be accurately concentric with the shaft, and the stationary component containing the pickups must be also accurately concentric. If the shaft moves radially, as for example an engine crankshaft is likely to do, then it may be necessary to connect the shaft encoder to the shaft via a flexible coupling.

3.3 Linear speed and velocity

Many of the position measurement techniques described above are used also for speed or velocity measurement by differentiating the signals with respect to time. In a similar fashion, velocity information is often derived by integration from accelerometer signals (described later in this chapter). In addition, there are certain techniques which measure linear speed or velocity directly.

Doppler effect devices measure the frequency difference between a transmitted and a reflected signal, and deduce the relative velocity from this. A well-known manifestation of this effect is the change in pitch of a siren on a vehicle as it goes past a stationary observer. Devices based on ultrasound and microwave beams are available.

Photographic methods can directly show the velocity of moving particles or markers (e.g. lights) on larger objects. If the time of exposure is known, the velocity can be found from the length and direction of the streaks on the photograph. This method is very useful for flow visualization, and is discussed in more detail under flow measurement, Section 6.3.3. There are particular types of errors which can occur with this method when focal plane shutters (as on a typical SLR camera) are used, and these are covered in the same section.

Another photographic principle is the time and distance method where a time exposure photograph is taken of a target marker under stroboscopic lighting. This can be a good method to reveal varying speeds.

Frequently it is convenient to mechanically convert linear motion to rotary motion, then measure the speed by one of the techniques in the next section.

3.4 Rotational speed and angular velocity

3.4.1 Basic methods

The familiar car speedometer uses a small magnet and an aluminium disc; the magnet is rotated by an arm, inducing eddy currents in the disc which is restrained by a hair-spring; the disc is deflected in proportion to the eddy current drag. The usual output is a pointer on a circular scale. Mechanical tachometers generally work on this principle or on a centrifugal principle.

Electrical tachometers (tachogenerators) use the dynamo principle: a rotating magnet is used as in the previous paragraph, but instead of generating eddy current drag the magnet induces a direct current (d.c.) voltage in the stator winding. If required this voltage can be fed back to a controller for closed-loop control.

Pulse counting methods are very commonly used, and appear in a number of forms. The optical tachometer detects pulses of light from discontinuities on a rotating shaft. These may be slots in a disc, a patch of white or reflective paint, or perhaps the blades on a propeller. Reflection from oil streaks have been known to give inadvertent extra signals. Magnetic equivalents using rotating magnets or slotted steel discs and inductive pick-ups are also commonly found. Inductive pick-up types can sometimes give double the number of expected pulses, due to the two flux changes at the leading and trailing edges of each slot. A particularly economical version is a rotating magnet actuating a reed switch, which is used in bicycle speedometers, hand-held cup anemometers and other relatively low-speed applications.

Pulse counting methods are applicable up to very high speeds, rotational speeds of the order of 3000 revolutions per second (180 000 rpm) having been measured accurately.

Stroboscopic methods, using a stroboscopic light or a mechanical stroboscope, are occasionally used. With stroboscopes there is a very real danger of gross errors caused by illuminating only every second or third revolution and under-indicating the speed by a factor of 2 or 3. If only part of the shaft circumference is visible, then it is also possible to over-indicate the speed by a factor of 2, due to the second image at 180° being invisible. One way of avoiding this is to use a helical marker around the shaft (if space is available), which will help in detecting double illumination. When using stroboscobic methods, great care is needed, preferably using a second indicating method as an approximate cross-check.

3.4.2 Shaft encoders

For high accuracy, shaft encoders are used, and these give both speed and/or position information; furthermore, unlike basic pulse counting methods, they can detect the direction of rotation when used with two signals 90° out of phase with each other. The direction is determined from which signal appears first. Readily available encoders may give up to 6000 pulses per revolution or operate to 12 000 rpm and above. They differ from the tachogenerators in giving a digital output, which may have to be converted to analogue for uses such as feedback control of motor speed.

An incremental optical encoder works in a similar way to the toothed wheel and slotted opto-switch, with a series of fine lines or grating on the rotating element, and light-emitting diodes to transmit and photodiodes or equivalent to receive the light. However, the light passes first through the rotating element and then through a second (stationary) grating to give diffraction fringes. This has the advantage that instead of the signal ramping from light to dark as each line passes, the whole field goes light or dark, so that a square wave is obtained.

It is the movement of these fringes that occurs when the encoder turns that is then sensed by the photodiodes. If electrical noise is a problem then opposing signals are available from sensors positioned in antiphase. Used with a difference amplifier, these can give an output sensitive to the signal itself, but relatively insensitive to noise, which affects both the signals. Sometimes a phototransistor may be used instead of a photodiode, to give an amplified signal at the encoder itself.

It is important that both the rotating and stationary part of the encoder are mounted concentrically to the rotating shaft or the fringes can be affected and measurements missed by the sensor. On a dynamometer with an optical encoder fitted, the encoder casing was mounted to the dynamometer on plastic spacers. The mounting screws became loose and the casing moved slightly relative to the shaft, giving incorrect readings varying from zero up to the actual shaft speed.

3.4.3 Gyroscopic devices

A gyroscope which is restrained (i.e. not gimbal mounted) and has its axis of rotation forcibly rotated, exerts a moment on its mounts proportional to the absolute rate of axis rotation. This moment can be measured using load cells, or used to directly deflect a needle against a spring force. This latter principle is used in a simple rate-of-turn indicator used in light aircraft. This method is obviously very valuable for angular velocity measurement where there is no readily available static reference.

The 'fibre-optic gyroscope' which incorporates a very long loop of optical fibre coiled up, with laser light sources and detectors, has been known for some time. Its principle of operation is that when two light beams propagate in opposite directions around a common path, they experience a relative phase shift depending upon the rotation rate of the plane of the path. This device therefore detects rotation rate, and the output can be integrated to give the absolute heading. These devices are of sufficiently low cost and high accuracy to be used in navigation systems for cars. Miniature low-cost 'gyroscopes' based on a vibrating beam principle are becoming available, and these too should be appropriate for this purpose.

3.5 Acceleration

Acceleration can be deduced by differentiation of a velocity signal, or double differentiation of a displacement signal. This applies equally to linear acceleration and angular acceleration.

Where acceleration is required to be measured directly, the most commonly used device is the piezo-electric accelerometer, which measures the force required to accelerate a small mass mounted within the device. The only significant problem with this type is the low-frequency response, so when signals of less than about 5 Hz are required to be measured accurately, it may be necessary to consider other types of sensor, for example ones based on strain gauges.

For measuring angular acceleration a pair of carefully matched accelerometers can be mounted parallel and a small distance apart, and their outputs subtracted.

3.6 Vibration measurement and interpretation

Vibration measurement can be considered as a sub-set of motion and displacement measurement. It is widely used as a technique for diagnosing machinery problems and for monitoring the 'health' of machinery, and therefore it is carried out in relatively uncontrolled conditions in the field, sometimes with inadequate understanding of how to ensure correct results, and how to interpret the results. Therefore it merits a section to describe the techniques and the pitfalls.

A simple steady sinusoidal vibration signal has a specific value of frequency, a specific value of displacement, and also values for velocity and acceleration which can be calculated from the amplitude and frequency. The values for displacement, velocity, and acceleration can be quoted as 'peak-to-peak', or 'amplitude' (= half of peak-to-peak), or root mean square (RMS). RMS is the most commonly quoted, but the other values are used, so when using an instrument or interpreting results, it is important to ascertain which is being used.

Mostly, vibration signals are not steady or sinusoidal, but a complete mixture of frequencies, often varying with time. For such a signal there is not a readily calculable relationship between displacement, velocity, and acceleration. In order to be able readily to compare vibration levels between different machines, it is common to use RMS vibration velocity over a broad band of frequencies.

The most commonly used sensor is an accelerometer (see Section 3.5). The output of this can be converted to velocity by integration of the signal, and to displacement by double integration. Basic instruments will usually give a reading of velocity, while more refined ones give a choice of acceleration, velocity or displacement. Care is needed with the units of measurement, particularly for acceleration which may be in m/s^2 or alternatively in '*g*'s, i.e. multiples of 9.81 m/s^2.

Other sensors are also used. For example proximity probes are used in plain bearings on turbomachinery to monitor relative movement and vibration between the shaft and the casing. In this case the output is displacement, which requires differentiation to convert to velocity or acceleration.

A British Standard [BS 7854, (**2**)] and various equivalent national and international standards (e.g. ISO 10816) define evaluation standards for vibration in various classes of machinery. These standards cover machine speeds from 600 to 12 000 rpm, and the vibration limits are based on RMS vibration velocity over a frequency band of 10 to 1000 Hz. Velocity is chosen rather than displacement or acceleration, because it has been found that by using vibration velocity, the acceptable levels are largely independent of machine speed.

The acceptable level depends on the type and size of the machine, and also on the nature of the mountings and foundations. The acceptable limits for a machine mounted on a flexible steel framework are substantially higher than for the same machine mounted on a solid concrete base. It is also worth noting that the user's subjective view of vibration level is often based on what he feels through the soles of his feet. Instances of reported bad vibration have been found to be resonances of steel floorplates adjacent to the machine, rather than problems with the machine itself.

The vibration measurements may vary substantially depending on the axis of measurement. The nature of machine mountings and foundations usually (but not always) means that vibration in horizontal directions is greater than in vertical directions. The standards are not totally clear on whether the vibration acceptance criteria should be different in a horizontal direction from in a vertical direction. Therefore it is important that the direction of measurement should be recorded so that subsequent interpretation is not confused.

Although vibration measurement can be quite precise, interpretation of the measurements is less so, and it is usually impossible to deduce very much about the health of a particular machine from a single set of measurements, even when comparing the readings with those from another similar or identical machine. This is because there are several factors which can cause significant differences in vibration readings – manufacturing tolerances, minor differences in mountings, minor differences in operational conditions, etc. The best way to interpret vibration measurements is by trend analysis. A progressive or a sudden change in vibration levels, if it cannot be traced to a change in operational conditions, is a strong indication that something is going wrong with the machine.

Spectral analysis of the vibration signal can be used to pinpoint specific faults. Specific frequencies can be identified as emanating from a particular gear, or bearing, or coupling, or turbine disc, etc. Again, trend analysis is much more valuable than single measurements, and a 'signature' trace taken from the machine when new is very useful when analysing later problems.

With spectrum analysers, it is important to ensure that the frequencies of interest are not being ignored or filtered out. During investigations of vibration problems with some large fans, it was found that the principal vibration with the largest displacement was at 5 to 10 Hz, while the spectrum analyser (in combination with particular sensors) effectively ignored everything below about 20 Hz.

Sometimes it is impossible to attach the vibration sensor directly to the point where measurement is desired, and a metal extension bar may be used. Such bars can have a number of lateral and longitudinal resonant frequencies which can easily confuse the measurements. It should be possible to identify these frequencies by applying the sensor and extension bar to a static machine and applying small impulses to the extension bar.

Rolling bearings can produce high-frequency vibrations with specific characteristics. There are special instruments developed to detect and analyse these vibration signatures, to determine the state of health of the bearing. As with general vibration measurements, single readings are not very precise, but analysis of trends on a particular bearing can be a useful guide to deterioration. Note that these instruments may not work reliably for low-speed bearings (under 500 rpm), and bearings adjacent to other sources of high-frequency vibrations (such as gears).

Case Study

Position sensing of photo masks using a semiconductor sensor

An inspection system was being developed to detect very small particles of dust or contaminants in the region of 150 nm diameter (well below the wavelength of visible light) on photo masks, prior to them being imaged on to silicon wafers. The process is shown in Fig. 3.2; ultraviolet light from a mercury lamp is converged by a condenser lens to give parallel light, which then shines through the photo mask.

The photo mask consists of glass with a very fine pattern of lines in chrome across it, which stops the light. The light which is not stopped by the chrome then continues through the imaging lens to the silicon wafer which is covered in a layer of material known as photo-resist, because it develops a chemical resistance by being exposed to light. The wafer is then inserted into a chemical bath which etches away the photo-resist where it has not been exposed, allowing conducting copper pathways to be added in its place.

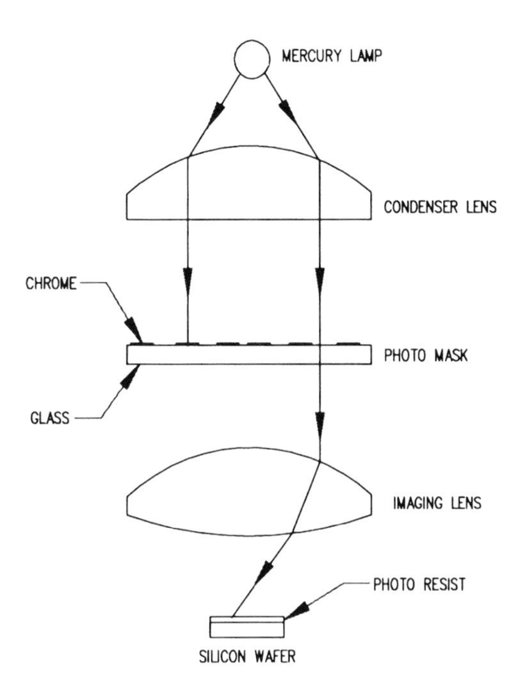

Fig. 3.2 The imaging process

The importance of the particles on the photo mask is that they scatter the light passing through, so that ultimately the copper pathways will be incorrect. As each photo mask is used to produce very large quantities of silicon chips, errors like this can lead to an extremely high rejection rate.

This led to the development of the inspection system shown in Fig. 3.3, in which a laser beam tracks rapidly across the surface of the photo mask. The reflected light from a flat surface is absorbed by a metal plate, whilst any particles on the surface of the mask scatter the light, which is then picked up by a microscope system with a line sensing camera. The image has then to be processed rapidly, and any contaminants identified. Unfortunately the height of the chrome on the glass surface is comparable with the size of the particles being measured, also causing diffraction of the laser beam and being picked up by the camera, and this has to be screened out during the image processing.

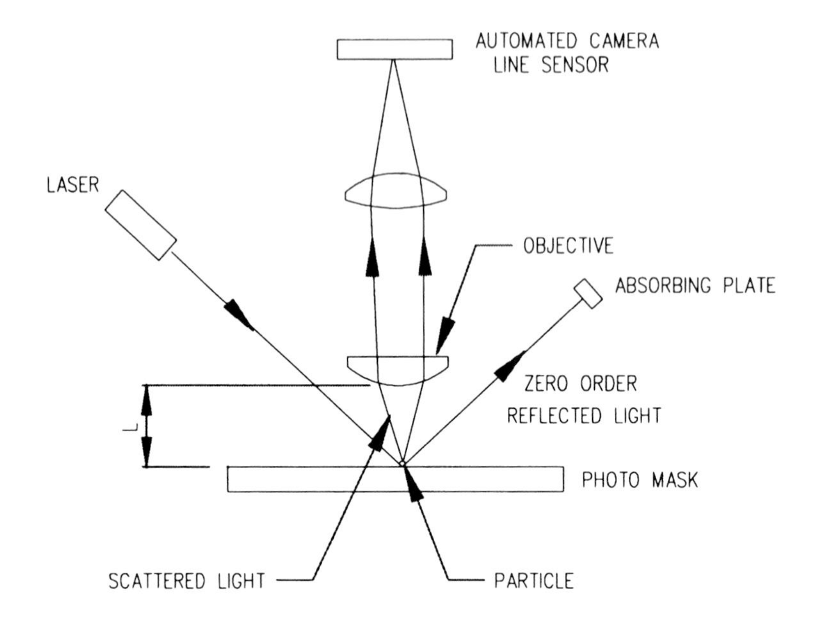

Fig. 3.3 Sensing contaminants on the photo mask

In order to keep the image in focus, the distance *L* between the photo mask and the objective lens had to be maintained to within about 250 nm during the tracking process. The photo mask was mounted on piezo-actuators (see Fig. 3.4) for fine levelling and for feedback control of position during tracking. Coarse positioning and levelling was done with manual screw adjustment, to within about 20 μm.

Height measurement had to be done using a method that was non-contacting, with a high resolution and a rapid response time. For this a silicon semiconductor position sensitive detector was used. This is not dissimilar in operation to the LVDT, in that it is energized externally, and has a split output to increase the responsivity of the instrument, and to normalize the result to make it insensitive to temperature changes, for example. However, in this case, instead of using a transformer and requiring physical movement, the transducer acts by light falling on the semiconductor surface and freeing electron-hole pairs which then migrate in the direction offering least resistance.

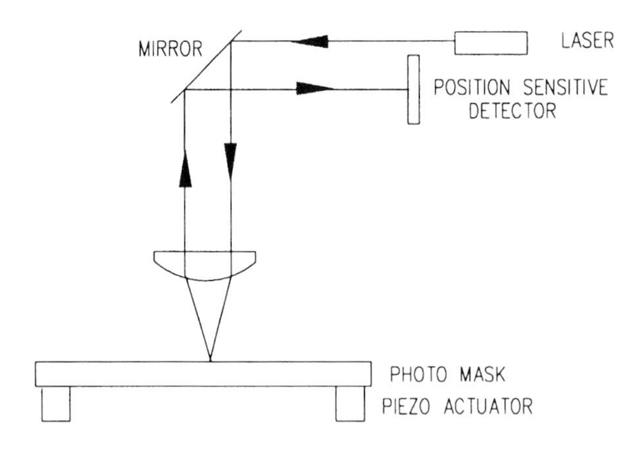

Fig. 3.4 Photo mask height measurement

The position sensitive detector was then checked for linearity over different laser intensities, and it was found that varying the laser intensity by a factor of 50 gave only 17% difference in signal output, which was extremely good. The detector was first calibrated using a micrometer to within 1 μm. However, this was the limit of the micrometer reading, and the detector resolution was well within this, so a LVDT had to be used for calibration purposes instead.

Case Study

Borehole probes

In quarrying, when drilling the boreholes in which to pack the explosive, it is important that the hole is neither too far from the blasting face, nor too close. If it is too far then the boulders resulting from the blast are too large to be useful, and if it is too close to the face then it can be dangerous, as rocks can then be blown great distances by the blast.

The position of the top of each hole is found by conventional surveying techniques. A laser mounted on a tripod and positioned in front of the quarry face is used, first to triangulate between two known points to determine its position, then to determine the three-dimensional profile of the existing quarry face, by positioning the laser at different angles and bouncing a beam off the surface, and calculating the distance by timing the reflected light. Then the position of the tops of each borehole must be found. As these will be out of sight from the quarry floor, a pole of known height is stuck into the borehole, with a bubble to ensure it is vertical, and a reflector to reflect back the laser beam.

The hole depth, incline and azimuth are then needed, assuming the hole is straight, to give an overall position of the hole relative to the face. Traditionally this is done by lowering a torch on a measuring tape to the bottom of the hole. The hole depth is read from the tape, the incline from sighting an inclinometer on the torch, and the azimuth by sighting a compass on the torch, which is difficult to do accurately.

However, if the hole is bent, or full of water, this method is not possible. A solution to this is the borehole probe, which can measure distance, inclination and azimuth at a series of positions along the length of the hole with reference to its top point, and without having to be seen from the hole top.

The construction of the probe is as shown in Fig. 3.5. Its core is two precision fluid level inclinometers mounted at right angles to one another, and an electronic magnetometer acting as a compass. It is self-powered using a rechargeable battery, and electronic circuitry mounted in the probe excites the instruments and converts the output into useful data.

This entire sensor package is mounted in a sealed stainless steel sheath, and is lowered by cable winch from a tripod into the borehole, on a flexible suspension cable. This also contains wires to transmit the data from the probe to a portable data recorder on the surface. An electronic cable length measuring device is used to determine the distance travelled down the hole.

The fluid level inclinometer consists of two circular discs which act as capacitors, milled out in the centre to contain the fluid. As the inclinometer tilts, the fluid level in the inclinometer changes, so that the dielectric constant between the two capacitor plates varies with the amount of liquid or gas connecting them. The transducer is energized with a pulse, and a pulse is returned, with the pulse width proportional to the angle of tilt. The pulse width is then measured digitally.

The magnetometer or flux gate compass consists of an energizing coil, pick-up coils and a ferrite ring. It operates inductively, with the field being

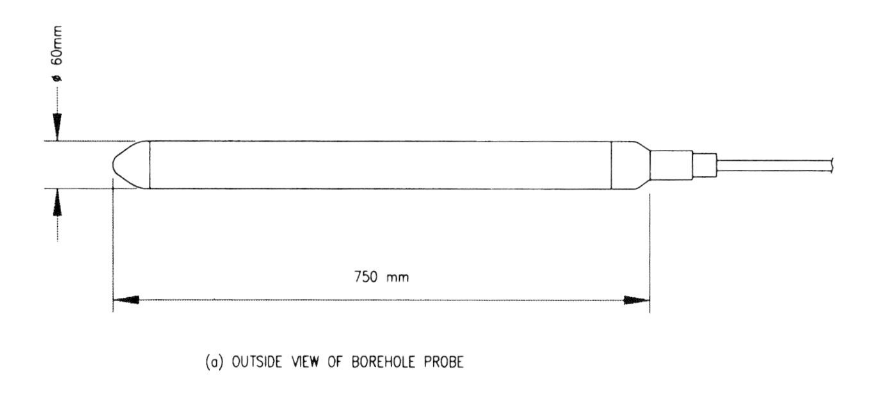

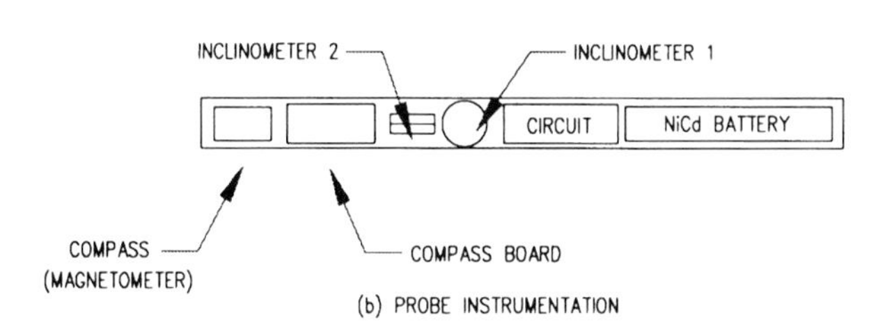

Fig. 3.5 Construction of the borehole probe

distorted by the earth's magnetic field. There are two pick-up coils, and the two output vectors are resolved to give the direction.

To measure total inclination, the inclinometers are mounted at right angles to one another and a microprocessor combines the two readings to give the absolute inclination of the probe. This is independent of the random rotation which occurs when lowering the probe into the hole. The outputs from the magnetometer are combined with the information from the inclinometers to measure the magnetic azimuth. Using these and the distance information from the cable length device, the position relative to the hole top can be obtained at intervals down the hole.

The survey starts at the bottom, then the probe is drawn up gently from station to station. It does not matter whether the probe rotates since each station gets an independent set of readings. The pulling-up tends to ensure that the probe stays in contact with the wall instead of hanging diagonally.

When taking a reading the computer refuses to accept it until the signals have become steady.

The linearity of the inclinometers up to 30° from vertical is within 0.5% or less than 0.15°. With a typical hole incline of 10° the azimuth accuracy will be within ±1°. Repeatability was found to be within 0.2 metres over a 20 metre hole. The only likely gross error in an electronic hole probe is when there is a significant amount of magnetic material in the rock, in which case the site is not suitable for this method of measurement.

The sensors used are able to withstand significant shock loading, such as being dropped on the ground, or even down the hole. In one case the probe was stuck in a hole, and the shot fired, then the probe recovered from the rock pile. Although the casing was wrecked, the sensors were still working and within calibration limits.

The software package which supports the instrument will plot profiles of the quarry face and the borehole, so that if necessary the holes can be re-drilled, or only part-packed with explosive. It can also be used to plan the hole positions and directions before drilling, working out the 'burden' or distance between the face and the explosives, and to calculate the rock yield.

Chapter 4

Force, Torque, Stress, and Pressure Measurement

Our standards of force are ultimately based on weight, magnified by levers as necessary. Sometimes direct application of weights is convenient, or we use spring balances which are easily checked by attaching weights. A minor problem with weights is that gravity varies perceptibly, mainly due to the earth's non-spherical shape. Gravity has been measured by free-falling body methods to be about 9.8127 Newtons per kg (or m/s^2) in London but would be about 9.7725 N/kg in Quito, near the equator and 2800 m above sea level [data from Kaye and Laby (**3**)]. Local gravity variations would not matter at all when weighing by comparing masses and only rarely in engineering work. It would affect the time-keeping of pendulum clocks, or could cause arguments about the price paid for a given mass of gold if weighed using a strain gauge device or spring balance.

This chapter is mainly devoted to measurements involving transducers. However, the simple mechanical methods still have a useful role. For example, spring balances may be spurned by the modern engineer or technician as being imprecise and incapable of being connected to a computer. However, they are cheap, portable, and do not drift unpredictably or go out of calibration. (Obviously, the spring may stretch or creep if permanently loaded, but this is readily checked by unloading.) They are also valuable for *in situ* approximate calibration of load cells, to check that the instrumentation as a whole is set up correctly.

4.1 Load cells

These devices are widely used for many purposes including overload warnings, tank content indication, measuring material quantities passing on a belt conveyor; if individually calibrated they can take the place of expensive mechanical weigh-bridges, which shows how much confidence they have earned. However, they require electrical instrumentation and may show a drift with time so that for critical purposes re-calibration may be required at intervals.

Most load cells use resistance strain gauges but there is no need to understand their detailed construction in order to use them. The main forms of load cell are the in-line design and the cantilever type known as a shear beam. In-line cells may be made for tension or for compression; many can be used for either, fitted with a domed button for compression, or hooks or other fixings for tensile applications. The loading need not be exactly central since the strain gauges can be arranged to compensate for some eccentricity (see below). In the absence of individual calibration the output is usually guaranteed only to 2% accuracy.

Commercially, cells are available for loads of many tons, others down to 1 g. For some applications the regular cells are inconvenient; devices such as those shown in Fig. 4.1 can be made and calibrated by dead-weights. Two precautions should be taken: firstly the electrical instrumentation should be as identical as possible for calibration and for the actual work, secondly the position of the load line may give non-linear performance, thus the calibration should cover the whole expected range, not depending on extrapolation from small loads or vice versa.

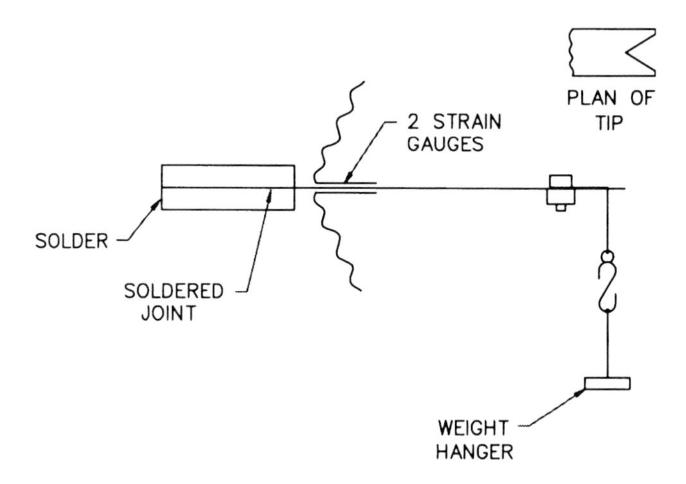

Fig. 4.1 Strain gauge set-up

Traditional platform scales have a lever system ensuring parallel motion, so that the potential energy change for a given deflection, and hence the reading, is the same at all load positions. When testing scales, the loads are placed at a variety of positions to check on this aspect. Scales using several load cells but no parallel motion mechanism should have matched cells with equal load–deflection properties and should have a highly linear output–load relation, otherwise the outputs will not add up correctly for all loads and load positions.

4.2 Metal foil strain gauges

4.2.1 Principles

The resistance of a strain gauge increases when put under tensile stress. The main reason is an intrinsic change in the material while under load (not a permanent change) but part of the effect is the increase in length and the reduction of cross-section as noted in Chapter 6 (the Poisson contraction). The effects are lumped together under the title of gauge factor. The gauges are bonded to the relevant surface; the meaning of the gauge factor is that if the gauge is extended by 1 part per million (one micro-strain or μe) and its gauge factor is 2, then its electrical resistance increases by 2 parts per million. If a gauge is attached to a surface with non-uniform strain, the resistance change relates to the average strain under the thin portion of the grid (the gauge length).

Thermal expansion of most structurally common metals exceeds 11 parts per million for each degree Celsius, whilst the gauge material could have a very different coefficient of expansion. Since working strains may often be below 100 parts per million, it is clear that small changes of temperature could give serious errors. The change of resistance with stress has been known for over a century but the temperature compensation problem was only overcome much later.

It is normally assumed that the gauge foil itself and the plastic carrier are so thin compared with the component to which they are bonded that it has the same elastic extension and the same actual thermal expansion (or contraction). Therefore unless its natural thermal expansion is identical with that of the component, the gauge goes into tension or compression as the temperature changes, even with no external load. The intrinsic resistance of the gauge material also tends to change with temperature. It has been found possible to select and treat metals used for strain gauges in such a way that the thermal expansion compensates for the resistivity change in the gauge material. The effect is called self-compensation; it can be made to suit typical steel values, aluminium or many other substances, for example concrete. In this connection one should note that, for example, stainless steels have a higher coefficient than mild steel, whilst high-nickel alloys range widely, including Invar with a low coefficient near 0 °C (though it has a much higher coefficient at high temperatures). Some aluminium alloys used for pistons have a low coefficient, others a coefficient higher than pure aluminium. A long-established supplier [(**4**), formerly under a different name] offers a great many levels of compensation. Nevertheless where the object is likely to vary considerably in temperature, the dummy gauge principle is adopted (see below). This also tends to equalize the heating effect which may come from the resistance-measuring instruments.

Modern strain gauges can stand extensions well beyond those commonly used in metal components, even the larger strains which may arise in plastics. It may be noted that in compression the resistance decreases, the gauge factor remaining the same over quite a large range. The permissible strains on some gauges are so large that they can be used to monitor crack growth and plastic extension.

The most common errors associated with strain gauges are not the gauges themselves being overstrained or damaged, but that they become unbonded from the surface to which they are attached. The readings will then drift or be unrepeatable.

Another problem with strain gauges is positioning relative to discontinuities. Often it is desired to measure the strains at a position with a high stress concentration such as a fillet radius. The exact positioning on the fillet is difficult to achieve consistently, and it is also difficult to maintain good bonding on a sharply curved surface. Therefore under these circumstances it is advisable to fit a second strain gauge nearby on a less curved surface where the conditions are better. The output from this second gauge then acts as a check on the condition of the first.

The second 'check' gauge also has the merit that it will probably be unaffected if a crack forms in the component in the vicinity of the first gauge. This effect was seen on a fatigue test rig, where the measured strain at a critical fillet radius decreased by about 30% as the test proceeded, while a second gauge about 10 mm away maintained a steady reading.

Investigation showed that the reason was the formation of a sub-surface crack on the fillet radius, close to the main strain gauge. This crack caused a re-distribution of strain, effectively casting a 'shadow' which reduced the strain reading from the gauge. On a test rig with defined loading, this effect was interesting but not critical. However, if using strain gauges to monitor the effects of random loads in service, for example on an aircraft wing, the effect could be dangerously misleading, indicating decreasing strains and loads when the component was actually cracking and close to failure.

Semi-conductor strain gauges are occasionally used instead of the metal foil gauges. These give a much higher output, e.g. 5–10 times that of the metal foil gauges, but are reported to be very temperature sensitive, and suffer from drift and hysteresis.

4.2.2 Construction of metal foil strain gauges

Figure 4.2 shows a typical foil strain gauge, though many have a greater length-to-width ratio. The delicate foil is attached to a plastic carrier (matrix). The shape of the grid is produced photochemically. There are two reasons for the large end-turns. One is that they provide secure attachment for the long thin grid bars, the other is that since the electrical resistance of the wide portions is very small, any change in resistance due to transverse strain can nearly always be ignored. BS 6888 1988 requires

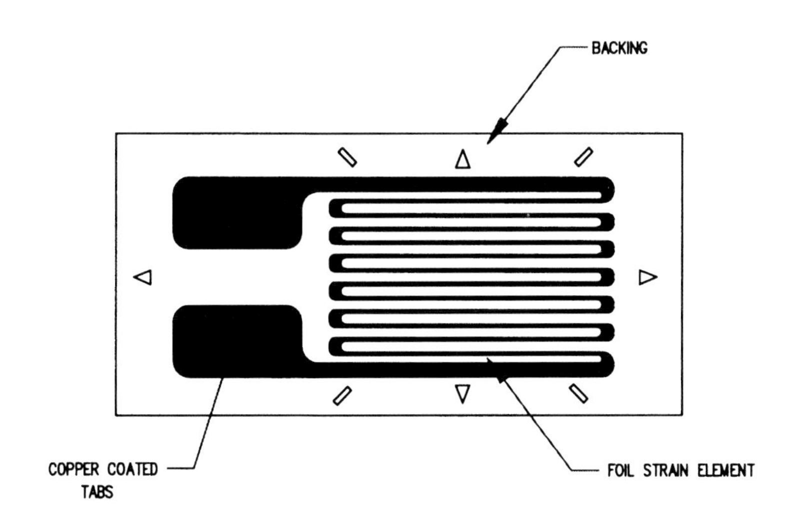

Fig. 4.2 A typical foil strain gauge

the transverse sensitivity to be below 0.15% of the linear sensitivity unless stated otherwise by the supplier. Some gauges are supplied with short leads attached, others provide anchor points prepared for easy soldering-on of leads. The top usually comes covered by a protective plastic film but can be obtained bare, which helps with heat dissipation but should at least be protected from moisture by a suitable varnish.

Strain gauges are normally available with an active length (gauge length) ranging from 2 to 13 mm. The shortest gauges are intended for non-uniform stresses. The finite length of the gauge itself can lead to some averaging, particularly when trying to pick up peak stresses around stress raisers. The overall length is greater than the active length, to ensure good bonding. There are also multiple patterns available to measure two strains at right angles, or three strains at 60° to each other. The multiple gauges are known as rosettes, the name being taken from one form which has three linear gauges superimposed. Many multi-directional arrangements are side-by-side rather than stacked, yet are still called rosettes.

The most usual resistance values provided are 120 ohms or 350 ohms, whilst the most usual gauge factors are 2.0 or 2.1.

Normally a gauge with its carrier is bonded to the surface of a component, which need not be flat though it should not be sharply curved. Double curvature calls for the use of very narrow versions and special attention during bonding, such as using pre-shaped resilient pressure pads to ensure that the bonding film is thin and uniform. For some applications it is convenient to purchase gauges ready bonded to a thin steel sub-base which can be attached to a component by spot-welding. This would apply mainly to flat or slightly convex surfaces, under tensile stress only.

4.2.3 Resistance measurement

The measurement may be by a balanced Wheatstone bridge (see Chapter 7), or by a bridge which is balanced at zero load and which measures the imbalance current in the galvanometer. Since the resistance changes are small it is not necessary to have a full-range variable accurately known resistor in the arm adjacent to the gauge, but to fit a fixed backing-off resistor equal to the gauge resistance, Fig. 4.3a. The arm may also contain a small, accurate variable resistor which is set, manually or automatically, to balance the bridge. This resistor gives the increase or decrease of the gauge resistance; it can be calibrated directly in micro-strains by setting up the gauge parameters. Alternatively, the backing-off arm is set up for a zero reading at no load and then kept constant so that the bridge becomes unbalanced under load, the galvanometer current being used to find the strain.

Instead of a backing-off resistor in the bridge box, it is often better to place a dummy gauge near the working gauge, mounted on similar material and located where the temperature is the same but not subject to load, Fig. 4.3b.

If a strain gauge amplifier is used the output may be given a zero offset to prevent it bottoming out if there is any drift. If the strain gauge is to be used bi-directionally, it is necessary to remember to remove this offset, or the output will be asymmetrical and incorrect.

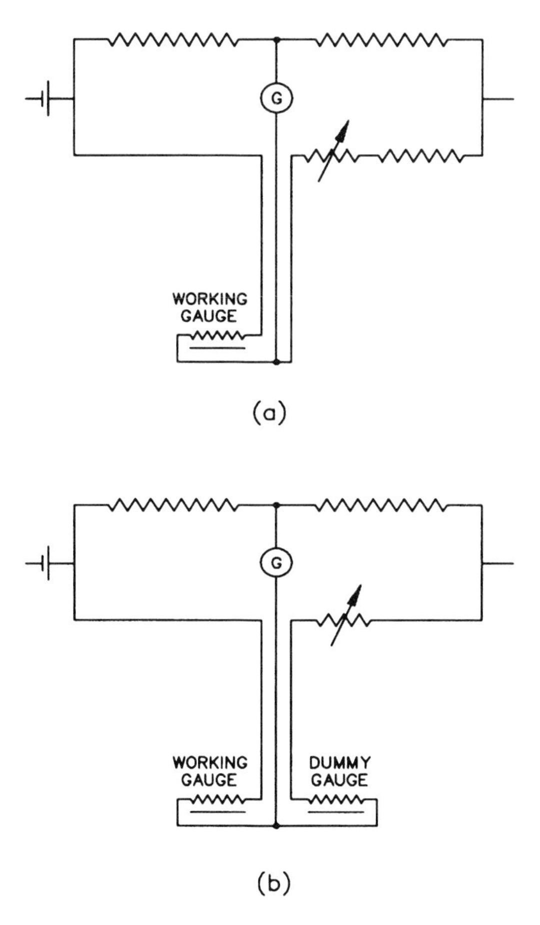

Fig. 4.3 Strain gauge resistance measurement

4.2.4 Leads resistance

Although the leads resistance may only be a fraction of an ohm, nevertheless it seems to be good practice to use the three-wire method for single gauges as well as for the dummy gauge method. This compensates out changes in the leads resistance in a similar way to the compensation of resistance thermometer leads, see Chapter 5. One of the two power leads is in series with the gauge, the other with the backing-off resistor or the dummy gauge in the adjacent arm. The third lead only carries the imbalance current. Alternatively a four-wire system can be used, corresponding to Fig. 5.6c.

The dummy gauge method is used when there are significant temperature changes at the gauge and/or there is uncertainty about the coefficient of expansion of the object, or worry about heating due to the instrument current.

It may be noted that these schemes still give an error, though it is usually insignificant: for example, if the gauge is of 120 ohms resistance and the leads each of 0.5 ohms (which could represent 50 m of copper cable, depending on size), then the fundamental output for a given strain is that of the gauges alone, but the reading obtained is only 120/121 of that value.

4.2.5 Bending effects

In cases of bending it is often possible to use one gauge on the tensile face, and another on the compressive face. Then there is no need for a dummy gauge since each resistance offsets the other in the unloaded state. Under bending, one resistance increases while the other diminishes, therefore the outputs are additive. If the component is thin it is worth remembering that the gauges are further from the neutral axis than the object's surface and therefore show a higher strain; it may also be that the stiffening effect of the gauges slightly reduces the deflection. These two effects work in opposing senses, reducing the error.

The converse of this principle is used to compensate for unwanted bending due to non-central loading. If we have a member in tension or compression, we can fit matched pairs of gauges on opposing faces and connect them in series. The pure tension or compression outputs are simply additive but any bending increases the resistance of some gauges while decreasing the resistance of their opposites by the same amount. The gauge factor of a gauge which satisfies BS 6888 has to be uniform to 0.5 % in tension and compression up to 1000 με (microstrain) unless stated otherwise by the supplier. It is also possible to put gauges into all four arms of the bridge to increase the output.

4.2.6 Excitation

The power for the bridge is known as the excitation, as opposed to any power to the indicating or recording equipment. It is always specified by the supplier, usually 10 V. It may be a.c., d.c., or pulsed. Alternating current causes heating in the fine grid bars and may also cause eddy current heating in the component or in the dummy gauge base; any temperature difference between gauge and base, component or dummy gauge defeats the temperature compensation. As heating of the gauge foil is likely to predominate, the makers recommend using gauges as large as possible. A bare-topped gauge shows a further advantage in this way but it is only slight, since moisture protection has to be provided.

With d.c. the grid heating is still present, though the eddy current problem is absent. On the other hand there is another problem, though we have not seen any reference to its severity. At the junction of the leads to the gauge metal the current produces a Peltier effect heating at one joint, cooling at the other, by the excitation current. This gives a thermocouple voltage which, it seems, must influence the output voltage at the meter. The pulsed form of excitation seems to be a way of reducing the Peltier heating problem. Presumably the instruments using this method provide a hold facility for the last reading and either a manual control or automatic ways of preventing needlessly frequent pulsing.

4.3 Torque measurement

For measurement of torque on static components, the most effective and simple method is to use a torque arm and a force transducer. Such an arrangement can be readily calibrated by means of weights. This is the method traditionally used for engine dynamometers, where the torque absorbed by the dynamometer is measured at the casing of the unit. If there are rapid torque fluctuations, or significant transient accelerations of the rotating components, this method can be inaccurate, and it is then preferable to measure the torque on the rotating shaft.

Torque transducers are available for mounting on rotating shafts. They are generally strain gauge or piezo-electric based, and transmit their output to the instrumentation either by slip rings or by radio signals. Use of such transducers is appropriate if they can be incorporated at the design stage, on a test rig for example. However, if torque measurements are required on an existing machine it is often physically impractical to incorporate a torque transducer, and even if space were available, the transducer may affect the dynamic characteristics of the machine. As the reason for carrying out the measurements is often to establish the dynamic torques and torsional vibration characteristics, use of torque transducers could give false indications.

The approach commonly adopted is to apply strain gauges directly to a portion of the rotating shaft, and to transmit the signals by a radio (or infra-red) link to the instrumentation. The equipment attached to the rotating shaft can be smaller than a matchbox, and has self-contained batteries suitable for some hours of operation. This method has been used successfully on a wide range of machinery in some adverse environments. If continuous operation is required, it may be necessary to provide a power input via slip rings or a transformer. Slip rings are sometimes used for transmission of the signal, but they are prone to noise, and therefore radio transmission is preferred where possible.

One potential source of error is the presence of undetected static torques. If the strain gauges are applied to the machine in service, then there may be a significant static torque applied to the shaft; possible reasons include friction in gears or bearings, or the use of automatic brakes which engage when the machine is stopped. The gauge will read the locked-in torque as zero, which can lead to quite large errors. A simple cross-check may be to measure the motor current under steady state running, and calculate the torque from data available from the motor manufacturer. Alternatively, it may be possible by manually rotating the machine slightly, to establish a position where the shaft is free of torque, and set the zero accurately.

Axial loads and bending moments can also significantly affect the reading. Ideally the torque should be measured directly at the point of interest, but this is not always possible. If there are couplings, gearboxes, or just substantial inertias or flexibilities between the transducer and the point where the torque is desired to be known, these can substantially modify the dynamic shape of the signal. The case study below illustrates this point.

Case Study

Problems with an in-line torque transducer

On a valve train wear rig used for testing of lubricants, an in-line torque transducer was used to measure the torque supplied to the cam-shaft via the drive belt. An unexpected increase of peak torque with speed was obtained, and when the results were studied in more detail it was found that the roughly sinusoidal signal corresponding to the valve lift had a higher frequency signal superimposed upon it. This affected the peak torques, giving unrepeatable results.

The cause was found to be frictional torque fluctuations in an Oldham type coupling downstream of the torque transducer. Modifications to the coupling were tried, but the ultimate solution was to use a bellows type coupling with no frictional effects.

4.4 Pressure gauges and pressure transducers

The familiar pressure gauges with dial and pointer (needle) mostly employ the Bourdon tube principle. A tube of oval cross-section bent into a C-shape reduces its curvature, i.e. straightens out slightly, when fluid pressure is applied. This happens because the outer and inner parts of the curve move further apart while remaining essentially unchanged in length. The movement of the free end is quite small and is magnified by some convenient means such as a tie-bar and lever or magnified further by a gear sector and a pinion on the pointer shaft, Fig. 4.4a. The method works under suction too, the tube radius becoming smaller. The zero is not normally adjustable, except by removing the pointer from its shaft and refitting in a different position. This lack of adjustment means that if the zero is found to have shifted, the gauge is likely to have been damaged or overloaded.

As the curved tube is almost impossible to clean out, the Bourdon type of gauge is not favoured in food, pharmaceutical, and many other processes. A diaphragm gauge is preferable, Fig. 4.4b. The fluid side of the diaphragm-type gauge is made cleanable by removing the gauge together with its diaphragm from the plant, leaving the mechanism undisturbed.

For remote indication and also for rapidly changing conditions pressure transducers are used; these have a diaphragm exposed to the fluid. The diaphragm may carry strain gauges, or press upon a piezo-electric crystal as described in Section 4.5. Alternatively, the deflection may be inferred from the capacity change between the diaphragm and a fixed electrode.

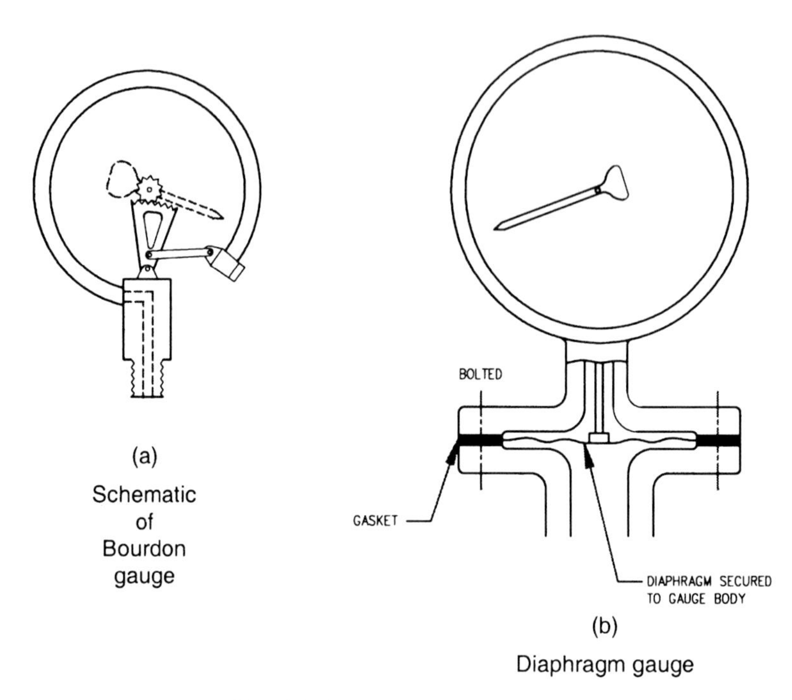

Fig. 4.4 Pressure gauge types

The strain gauge type uses a Wheatstone bridge instrument as described earlier; under steady conditions it could perhaps be used with a manual Wheatstone bridge to measure the resistance change under pressure, though a fixed bridge is more convenient; this becomes unbalanced by the resistance change, the unbalance being used for the pressure indication. The other types require electronic circuits to create a signal suitable for transmission to an indicating or recording instrument. To have confidence in the readings it is best to have a static instrument (gauge or manometer) either permanently connected to the same service, or ready to connect by opening a valve, so that a static comparison may be made of the transducer plus its instrumentation. The dynamic problems are discussed below.

Most transducers are made insensitive to temperature changes by choosing materials with appropriate coefficients of thermal expansion. In this way the resistance, the capacity, or the force on the crystal can be made independent of slow temperature changes. However, there is a serious difficulty with rapid temperature changes. Supposing that the fluid temperature increases quickly, the exposed face of the diaphragm expands first so that the diaphragm bows outwards. This has the same effect on the signal as a reduction of pressure. When measuring fluctuating gas pressures there is normally a temperature change accompanying each pressure change, hence false readings are quite common. An extreme case is the measurement of pressures in the combustion chamber of an IC engine, where there are very large and rapid temperature fluctuations. In this application, some pressure transducers can give dramatically incorrect readings. The error may be reduced if some non-rigid thermal shielding can be placed over the diaphragm, which hinders heat flow but allows the pressure change to reach the diaphragm. Coatings of silicone rubber have been used to good effect. It may be thought that Invar alloy could be used for the diaphragm in some cases. Invar has a very low coefficient of thermal expansion over a certain range of temperature.

If a pressure transducer is to respond to rapid changes in a gas, an obvious precaution is to avoid creating a resonance chamber. Unfortunately many pressure transducers are constructed like a Helmholtz resonator, with a reduced passage leading to a chamber in front of the diaphragm. If a volume V is connected to the pressure region by a duct of cross-sectional area A, length L, then the natural frequency is readily derivable from first principles, being given by

$$\frac{1}{2\pi\sqrt{\gamma\,.PA\,/\,\rho V L_{\mathrm{E}}}} \qquad \text{(Hz)}$$

where γ is the adiabatic index of the gas, P is the absolute gas pressure, ρ is the gas density (note that $(\gamma.P/\rho)^{0.5}$ is the speed of sound in the gas concerned, at the prevailing temperature). L_{E} is the equivalent length of the passage allowing for end effects. An approximate figure for L_{E} for a round hole is the actual length plus one diameter. A similar expression in another form for slots is given by Turner and Pretlove (**5**). Their expression should also apply to liquids where the cavity frequency is much higher than in gases provided the cavity is really full of liquid.

In liquids the installation should be such that no gas bubbles can accumulate. As it seems to be customary to place a transducer either vertically above a pipe or vessel, or horizontally, the chamber in front of the diaphragm is almost certain to hold air bubbles For fast-changing conditions the transducer diaphragm should be as nearly flush with the surface as feasible and installed to allow gas bubbles to rise away, and preferably not on top.

4.5 Piezo-electric transducers for force, pressure, and acceleration measurement

These transducers are relatively expensive, and their charge amplifiers even more so, but they have advantages in a number of situations. They are active transducers, utilizing the piezo-electric effect by which if the crystal lattice of, say, a quartz crystal is deformed under load it generates an electric charge. If a number of the quartz plates are then loaded in series and connected in parallel this increases the magnitude of the charge, which can then be fed through a charge amplifier and measured without serious leakage. Piezo transducers are always used with a charge amplifier to oppose the leakage, but nevertheless they are intended for dynamic conditions and not for static loads or pressures.

Between the transducer and the charge amplifier a special high-insulation, low-noise cable is used. This is manufactured under highly controlled conditions and in a vacuum to avoid oxidation at the joints. It is not a good idea to remove the end connector and then shorten the cable and re-solder the joint, as this can lead to severe loss of signal at this point! Downstream of the charge amplifier any cable can be used.

Piezo transducers are commercially available for dynamic force, pressure, and acceleration measurement. They were first used in IC engine development, measuring the cylinder head pressure, with the advantages of compactness, high natural frequency and a low temperature sensitivity. They are not particularly susceptible to acceleration, but for measuring low pressures under severely vibrating conditions it may help to use a low pass filter to remove the vibration signals, or resort to an acceleration-compensated version. Normally extensive temperature compensation is used, so that the temperature error is very small, for static or slowly varying temperatures. As well as the quartz crystals, both silicon and ceramic piezo-electric sensors are now available, with the ceramic sensors having a wider temperature range and much higher sensitivity than quartz.

In force measurement the piezo-electric sensors have the advantage of high spring stiffness so that there is almost zero displacement and an excellent frequency response. If high stiffness is critical it may make the difference between choosing a piezo transducer or a load beam. They can be constructed to measure in one to three planes; the 'cross-talk' between the different planes caused by manufacturing limitations being normally insignificant. However, in one case where low friction coefficients were to be measured a 1% cross-talk from the normal load would have swamped the friction reading. Their high stiffness and the possibility of cross-talk mean it is particularly important to de-couple any loads you do not wish to measure upstream of the transducer.

For example, in a test machine simulating the loading on artificial hip and knee joints, the transducer was to be used to measure the normal loading and the hydrodynamic friction in the joint during movement. Ultimately two transducers were used; a load washer in line with the normal load, and a transducer at 90° to the load, de-coupled with flexures, to measure the friction. The load washer was pre-loaded, avoiding any tensile stresses due to bending moments, with an elastic-shank type bolt to reduce the loss of sensitivity due to loads being transferred through the bolt.

Attention has to be paid to the mounting of load washers. For example, the supporting and mounting surfaces should be level and rigid (see Fig. 4.5a) to give a uniform stress distribution across the washer. Figure 4.5b shows deflection of the pre-load screw loading the inner edges of the washer to give an incorrect output value. Insulating washers can also be used if earthing presents a problem.

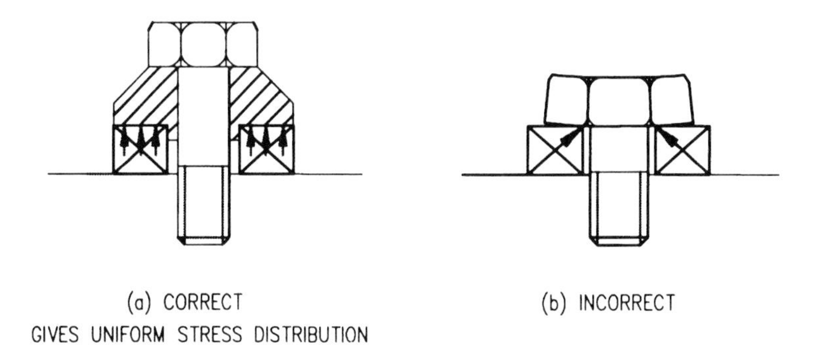

Fig. 4.5 Method of mounting load washers

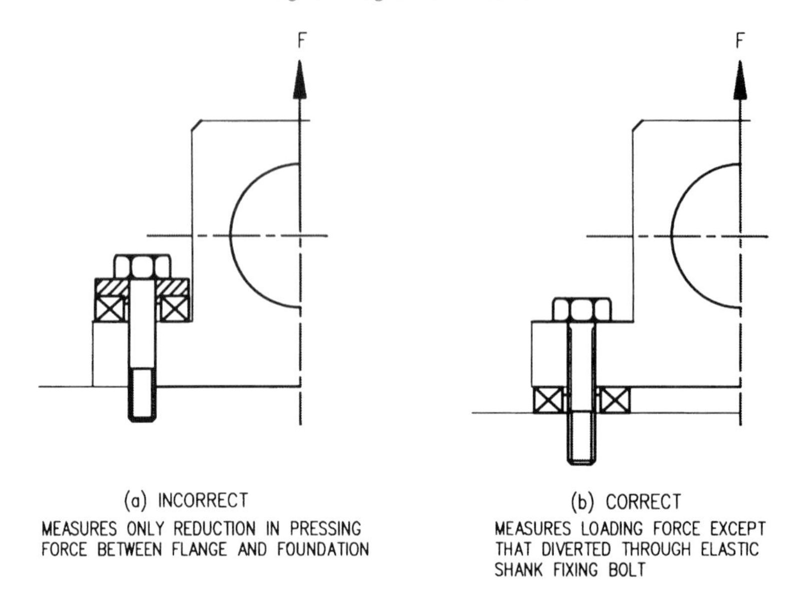

Fig. 4.6 Position of mounting load washers

The position of mounting the washer is also important; Fig. 4.6a shows a load washer being incorrectly used to measure a bearing force, as it measures only the reduction in pressing force between the flange and the foundation, not an increase in the screw force. Figure 4.6b shows the correct mounting arrangement, so that the load washer sees all of the force except that diverted through the fixing bolt. It is calibrated *in situ* to determine this. Further details of the correct use of the transducers are given in reference (**6**).

4.6 Optical and fibre-optic based sensors

In recent years a number of sensors for force, pressure, etc. have been developed based on optical devices (light or infra-red), communicating their signals via optical fibres. These have several merits, the principal one being complete immunity to electromagnetic interference. Other advantages include inherently high-frequency response, and the ability to pass signals across small air gaps or through windows. These are not necessarily exotic or expensive devices, and are appearing in various applications including automotive. These devices are effectively an extension of the principle of the opto-isolator, a well-known and economic device for transmitting electrical signals while maintaining electrical isolation.

4.7 Manometers

For low pressure differences manometers are widely used, using the levels of a liquid to give the pressure difference. The chief problems lie in the lines which connect the manometer to the source of pressure; since a major use is with flow measurement devices, this aspect has been covered at some length in Chapter 6; however, it applies to manometer use in general.

At the manometer itself, the first point of attention is the scale, which may be a direct measurement or may be scaled so that only one reading is needed, with compensation for level change in the other branch (usually an enlarged reservoir). The second point is the liquid used. In many instances the scale is intended for water but some manometers use a mixed fluid which has more consistent properties of wall-wetting, to minimize surface tension problems. For some applications with large pressure differences, mercury is used; if so one must watch out for an oxidized skin which may affect readings. The laboratory may have an apparatus for filtering mercury. Mercury vapour is toxic, hence only properly enclosed equipment should be used.

A sensitive manometer useful for calibration is the Casella instrument. This uses mercury, with a fairly large surface, perhaps 20 mm in diameter, to cut down meniscus curvature errors. It is fitted with a microscope and the observation is made accurate by using a conical point on a micrometer screw; the reflection of this point in the mercury pool allows a sensitive detection of the level. It is important to stand the instrument on a very firm table and check its position by the inbuilt spirit level.

Manometer readings may be magnified by using a sloping tube, in which case the above remarks about a firm table and levelling are particularly important. Another magnification method is to use two non-mixing liquids in a differential mode, so that the pressure difference is the level change times the difference between the two densities. This is a little cumbersome since both densities need to be known accurately. Industrial flow recorders are available which include a manometer for low differentials, in a chamber strong enough for high pressure. The manometer uses mercury, the height change being transmitted by a float operating a lever system or an electrical displacement transducer (thus avoiding a moving seal problem). Another method does not use a float but detects the mercury level by carbon rods dipping into the mercury so that their resistance decreases as the mercury rises.

4.8 Floating bell instruments

A brief mention is justified of floating bell instruments, used chiefly for chimney draught or ventilating systems. They can conveniently carry clockwork chart recorders. The action is by opposing the pressure above or below ambient against the buoyancy of a hollow metal can and appropriate counterweights, partly immersed in water. The bell rises or falls with the measured pressure.

4.9 Method for measuring small near-horizontal forces

The advantages of this method are absence of electrical apparatus, absence of pulley friction and low cost. Figure 4.7 shows a stand carrying a thread, 0.5 to 1 metre long. This carries a simple weight carrier easily produced from wire and plastic or aluminium, a ruler and another thread leading to the point of interest. The application in this case was measuring the drag of objects in slow-moving water. The force is found from the weight carrier plus added weights. For improved accuracy, allowance can be made for part-weight of the threads, and corrections for the slope of the near-horizontal thread (downward component of drag force added to the applied weight and cosine correction for the drag). This very simple apparatus has been used to measure forces of a few milli-Newtons in a virtually self-calibrating manner.

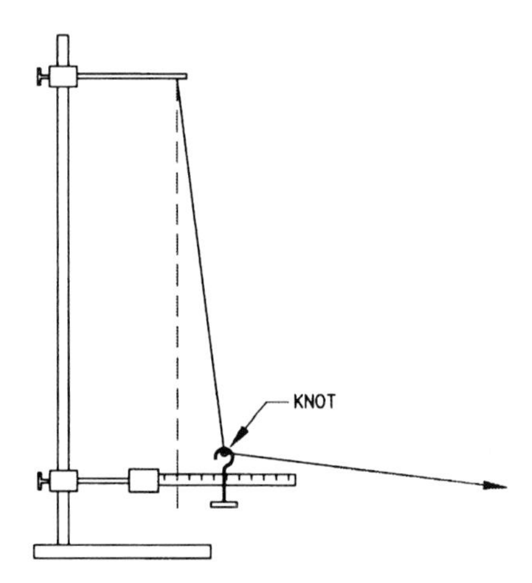

Fig. 4.7 Method for measuring small near-horizontal forces

Chapter 5

Temperature Measurement

This chapter deals mainly with glass thermometers, metal thermocouples, and resistance thermometers. Surface temperature measurement is mentioned, involving both contacting and non-contacting methods. Thermistors, though widely useful for their choice in temperature-versus-resistance properties, are not generally marketed for routine temperature measurement. This is probably because their characteristics are rather non-linear and their resistance can be very high, requiring more advanced instruments than thermocouples and metal resistance thermometers. They are, however, commonly used for temperature control, for example in refrigeration plants. Semiconductor temperature sensors are available within a limited temperature range, but are not, at the time of writing, widely used.

One or two considerations when measuring temperature, flagged up by BS1041 (7), are that firstly the contact between the measuring body and the substance being measured should be satisfactory, and secondly that the testing body should be small enough not to disturb the temperature conditions or cause too much thermal lag during transient testing. Another point is that there should be no chemical reactions between the testing body and the substance which will cause heat to be produced or absorbed. Where there are temperature gradients it is advised to test along an isothermal, and one should avoid condensation on the testing body as the latent heat of condensation will affect the temperature reading.

5.1 Glass thermometers

The well-known mercury in glass thermometer is rightly regarded as vital to the practice of temperature measurement. It should be generally known, however, that it has sources of error, connected with the use of glass: the first is sidereal or long-term error. It is mainly a zero drift and simply needs a note attaching to the thermometer stating the last check date and the extent. The cause is that the glass bulb creeps slowly under its own surface tension, or internal pressure if any, hence its volume changes over a time.

Measuring the zero error is not quite as easy as one might imagine. Often calibration of thermocouples, thermometers, etc. is done by inserting into a solid heated block, using a reference thermocouple, which depends on adequate contact and the reference thermocouple being correct. Another method is to use a fluidized bed of sand. Hot air is blown through the sand, effectively turning it into a fluid for good heat transfer. This is particularly useful for high temperatures.

Otherwise methods such as melting ice in a Dewar (Thermos-type) flask can be used as shown in Fig. 5.1. However, this accumulates water in the bottom; water is anomalous among liquids, being densest some way above freezing point, at about 4 °C. Thus a thermometer with its bulb in such a vessel may truly be measuring something between 0 and 4°C. This is easily overcome by extracting the water frequently with a plastic tube and suction bulb. It is also possible to buy a delicate device, requiring slight refrigeration, which tests thermometers at 0.01 °C, the triple point of liquid water, ice, and vapour.

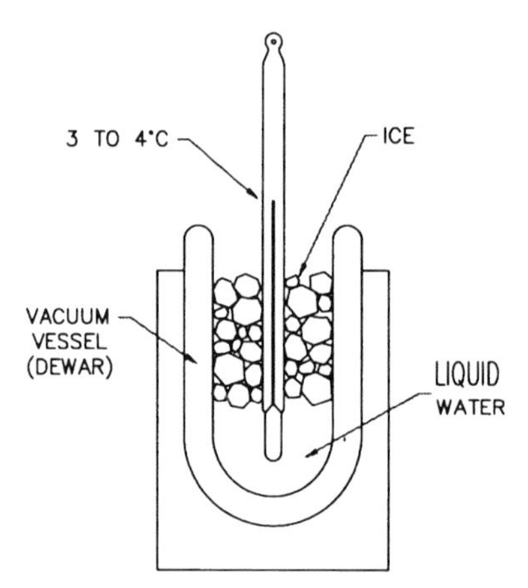

Fig. 5.1 Temperature calibration using melting ice

A thermometer can also be checked at higher temperatures; boiling water is not ideal since water requires some superheat to overcome surface tension before it can form the bubbles of steam. The use of fragments of tile reduces the extent of superheat but nevertheless it is preferable to use a condensing vessel (hypsometer), in which the thermometer is totally immersed, possibly involving stem corrections as described in BS 1041 Part 2.1 (7). The boiling or condensing point has to be corrected according to the barometric pressure; the instrument was originally intended as an altimeter for mountaineers, to find altitude by referring to tables of pressure versus boiling point.

For higher temperatures the freezing points of various pure substances are used, with the thermometer in a block of copper or aluminium placed in the molten substance, with a suitable oil around the thermometer to aid thermal contact. Freezing points are more definite than melting points because during melting it is difficult to ensure uniform temperatures. As the molten substance cools in the crucible the temperature reaches an arrest point as the latent heat is liberated while most of it is still liquid.

The other error is immersion error; this will rarely exceed 1 °C. The immersion depth at calibration is marked on the thermometer, usually 75 or 76 mm (3 in), though some are intended for total immersion. Also, the exposed liquid column of mercury is rarely at the bulb temperature but somewhat cooler. If this temperature can be estimated or measured, the error due to the exposed mercury column can be found using data in BS 1041.

If there is a great risk of breakage, spirit- or glycol-filled thermometers may be preferred since these liquids have less toxic effects than mercury vapour.

5.2 Dial thermometers

There are many varieties of dial thermometer based on bimetallic strips coiled up in spirals or helices; some just show room temperature, others reach a few inches into compartments like drying ovens or cool stores. Bimetallic strips use the difference in thermal expansion between two metals bonded together; they are also widely used in thermostats where they tend to show a drift towards hot (late opening of contacts with rising temperature). The reason is not clear; it may be due to inter-metallic diffusion, which might also affect the bimetal in thermometers. If that should be found true, a trimetal strip with a thin mid-layer might be useful.

When a longer reach is required a more expensive form is used, with a bulb, capillary pipe, and a pressure-gauge-type mechanism. This type may be liquid-filled, actuating the gauge by direct thermal expansion of the liquid in the bulb. The volume of the capillary pipe is small but gives an error according to its mean temperature. A partly filled system may be used relying on vapour pressure of a suitable substance. These types are nonrepairable but quite tough against minor accidents. The liquid-filled type has the advantage of failing instantly whilst a vapour-pressure type can leak for a time giving a good reading at first, then fading slowly, which may deceive the user.

5.3 Thermocouples

5.3.1 General principle

If two different metals A and B are connected as in Fig. 5.2, with a junction and a voltmeter, then if the junction is heated the meter should show a voltage. This is known as the Seebeck or thermocouple effect. It is part of a broader picture since in an electrical conductor a difference of temperature is associated with a difference of electric potential. The potential varies in different materials.

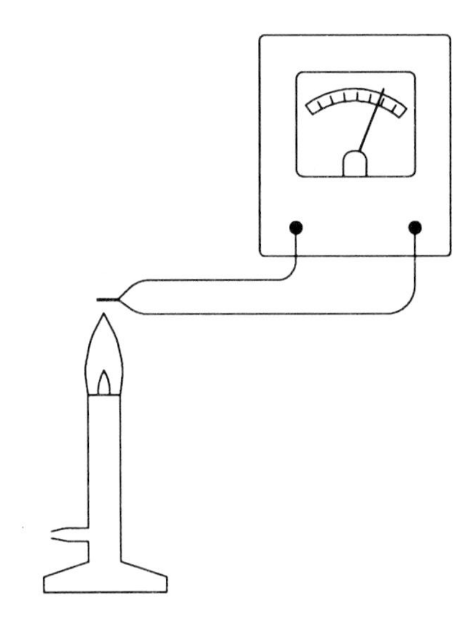

Fig. 5.2 The Seebeck or thermocouple effect

5.3.2 Materials used

A number of pairs of metals have become recognized as useful and are made up to a closely controlled composition. A couple made to the requisite specification will give an electromotive force (emf) as stated in BS EN 60584-1 (**8**) or other published tables, without need for individual calibration. The level of accuracy is also defined in standards.

As a first guide, the output in microvolts per °C is around 41 for the K type (chromel/alumel, or T1/T2), around 50 for the J type (iron/constantan), while the platinum-based couples range around 10. The latter are not startlingly expensive, since only fine wires are involved. The choice is made mainly on the basis of the temperature range. The most widely used couple is the chromel/alumel type up to 1000 °C. It is slightly affected by reducing atmospheres, presumably due to the aluminium content. For these cases of atmosphere the iron/constantan (J) type is more resistant, but is only recommended up to 800 °C. For higher temperatures the N type has been introduced, based on nickel/chromium/silicon versus nickel/silicon. The copper/constantan (T) type, though said to be favoured for sub-zero temperatures, has the disadvantage of larger heat conduction errors (discussed below).

In the aerospace industry N-type thermocouples are often used around 1000 °C due to their high-temperature stability. It is reported that K type couples drift by as much as 5°C at this temperature.

For higher temperatures and possibly difficult atmospheres there are several standard combinations of platinum and platinum/rhodium alloy. Furthermore there are some non-standard couples available using tungsten versus rhenium or iridium/rhodium versus iridium.

Considering the variety of pairs available it would be too difficult to include a fair selection of emf tables in this book. The emf of chromel/alumel is found in many handbooks and fuller tables are available from suppliers. The emf of any listed couple is enough to feed present-day instruments and signal conditioners.

In looking at emf tables it will be noted that the voltage is quoted referred to 0°C for convenience. There is no need for either junction to be at 0°C; the emf of any couple is simply the difference between the values for the junction temperatures. The temperature required is usually the hot junction but the case applies equally to low temperature measurements; for consistency's sake the description adopted here is the sensing junction and the reference junction.

5.3.3 Connection to instruments

In most cases there is no deliberate reference junction, it simply occurs where the couple is joined to a common metal at a common temperature. In laboratory work the thermocouple may often be taken right through to the instrument, or to a connecting block at room temperature from which copper leads are taken to the instrument. The effect of using a voltmeter or other instrument must be taken into account since there will be a loss due to resistance of the circuit unless a potentiometric (null) method is used (but see below).

Alternatively, thermocouples can be obtained with a signal processing circuit (transmitter) connected to the end of the couple itself, which gives an output ready for remote instruments such as digital indicators or data-loggers. Outputs may be 4 to 20 mA, 1 to 10 V or similar ranges, whilst presumably avoiding the flow of appreciable currents in the thermocouple. The energy for these is supplied from a constant low-voltage source. It is important to follow the instructions since the output does not necessarily start at zero but tends to have an offset above zero to suit the recording instruments. In deciding whether to use a transmitter mounted directly on the end of the thermocouple, attention is required as to whether there is danger of excessive temperature at that point which could affect the electronics.

In cases where the ambient temperature of the reference junction could fluctuate excessively for the purpose in hand a 'hot box' can be used. This is a block heated by an independent power supply and governed by a thermostat. The only drawback is the chance of thermostat malfunction or the box being switched off inadvertently. An alternative is a more massive box but without power input which is temperature-monitored and could be recorded on a spare channel and even made automatically to correct the reading of the sensing junction.

In important cases some thought should be given to the following: when a current flows through the couple there is a Peltier effect which heats or cools the junction. Along the wires there is also heating or cooling; this is due to the Thomson effect as the current flows in the wires with or against the temperature gradient. There is also resistance heating due to current flow, particularly if very fine wires are used. Whatever method of measurement is used, even with a potentiometric null method, some current flows either all the time or during the balancing process. Some instruments re-balance many times per second. We have not seen references to the magnitude of these effects but when great accuracy is important it may be wise to set up a test using the thermocouple and instrument, in relevant conditions, in a chamber with a mercury thermometer for comparison.

In industry, is not usual to run the thermocouple itself to a distant instrument but to use an extension cable. This may also apply if a transmitter is used but has to be kept away from hot equipment. Extension cable is made of the same material as the couple but with cheaper, tougher insulation, and possibly thicker wire for lower resistance. Then if the couple suffers from heat or other damage only a new short couple need be fitted. A junction of identical materials does not affect the emf. In cases of long runs, the extension cable may in turn be connected to a compensating cable, at some point where the temperature is closer to room temperature. Compensating cable is generally of different material, of lower resistance or lower cost, but has a matching emf at the temperature where the connection is made, so that the overall emf is unchanged. Extension and compensating cables may be subject to electrical interference, so often signal conditioning is done close to the instrument and only the processed signal has to be transmitted.

Galvanometers or potentiometers calibrated not as voltmeters but directly in terms of temperature for the relevant couple are usually fitted with cold junction compensation, so that the reading refers directly to the required (hot) junction temperature. Moreover if the instrument is a galvanometer it may also be already corrected to allow for a typical couple resistance, stated on the dial, for example 15 ohms, whilst the meter resistance will be much more, perhaps 2000 ohms. If the circuit resistance is different from the assumed value there is not much difficulty in making a correction taking the ratios of the circuit resistances, including the meter. A greater resistance results in a lower reading for a given actual temperature and vice versa. The correction is usually quite small. Manual potentiometric voltmeters do not require a resistance correction but require a knowledge of the effective cold junction temperature, usually the instrument temperature as noted above. The same is generally true of electronic instruments of high impedance, digital or dial-type.

5.3.4 Thermocouple availability

Thermocouples can be bought ready-made in various convenient forms of shielding etc. at very reasonable cost, but if many are required they can be made up from wire of the correct specification. The J type junctions and probably many others can be welded with a small oxy-acetylene flame. Other methods are capacitor discharge, or purpose-made commercial micro-welders. The wire may have insulation such as glass fibre but in any case the surface tends to be oxidized so that the intended junction should be cleaned at the tips and twisted together for support over a short distance, before welding. Further back the wires, if bare, can be supported and separated by twin-bore refractory sleeves, with short beads where bends are needed.

Numbers of couples can be welded or otherwise joined to a common base, e.g. for examining temperature distribution on some object, as an alternative to thermographic pictures. Each couple output should be measured separately. Using single wires on a common base could be affected by the emf due to temperature gradient.

Thermocouples should preferably not be immersed in water without some varnish or other protection since under some conditions a very powerful electrolytic voltage may develop, possibly larger that the thermal emf.

Mineral insulated, sheathed thermocouples are commercially available for use in liquids and solids with a reasonable conductivity; the sheath protects and insulates the thermocouple wires. The sheaths are commonly supplied in stainless steel or Inconel. A thin layer of the insulating material contains the thermocouple wires to prevent them earthing against the sheath. These thermocouples can readily be bent to the required shape.

The insulation will give some loss of sensitivity, particularly to transients; sheathed thermocouples are also available with a junction earthed to the sheath, giving a faster response time. (See also Section 7.2.2.) For use in air or other gases thermocouples are available with a non-contacting sheath with holes in it where the gas passes over the thermocouple itself. The sheath in this case merely locates and protects the thermocouple wires. Such an exposed couple must not be used if a chemical reaction, or catalytic action, could take place.

5.3.5 Heat loss errors

There are two slightly different problems here. In many cases the couple (or even a thermometer) has to be placed in a pocket (sheath, sleeve) either to protect it from the fluid or to enclose the fluid, which may be cooling water, exhaust gas, cooling air, lubricating oil, compressed air, or some other process substance. This makes it easy to inspect or change the thermocouple (or other sensor) without stopping the process. Sometimes very simple pockets are used and can cause enormous errors; Fig. 5.3 shows some examples, in cases of air flow at atmospheric pressure. In liquids the effect is much less serious. However, for measuring stationary liquid temperatures it is important that the thermal mixing is adequate.

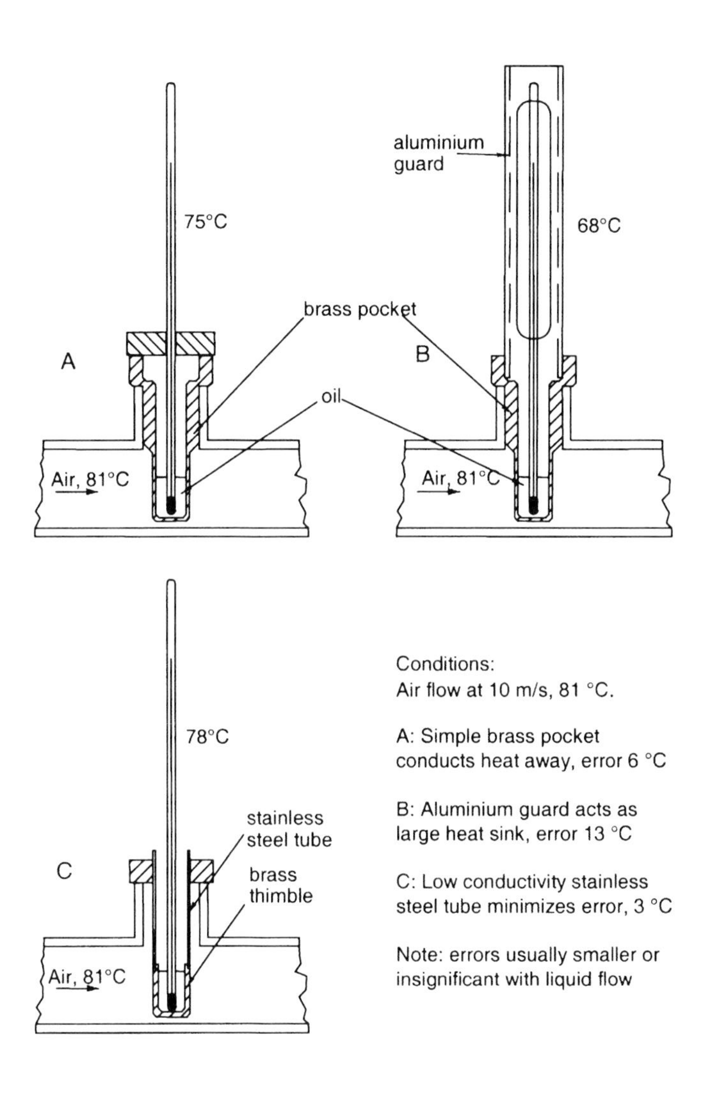

Fig. 5.3 Examples of errors caused by different pockets

The immediate environment of a thermocouple is usually an air space, or an air or gas stream. The temperature attained by the junction depends on the heat transfer from the air to the couple; at the same time heat is conducted away along the wires to the atmosphere. The still air space in a pocket is not a good heat transfer agent, thus it can make a large difference whether the couple is free or held firmly against the pocket wall at all times. Often the thermocouple is spring-loaded against the pocket wall. Sometimes it is feasible to bend the couple back a short distance to offset the temperature gradient, particularly in temporary rigs rather than in industrial installations.

To give some impression of these losses, in some experiments with couples directly placed in a good air-stream (2.5 m/s) it was found that a bare chromel/alumel couple of fairly fine wire or a purchased thermocouple with the junction welded to a thin stainless steel sleeve (for good thermal contact) had an error of only about 1%, if immersed to a depth of 20 mm, since chromel and alumel and stainless steel are poor heat conductors. In still conditions much more immersion is needed to keep down the conduction error. For commercially available sheathed thermocouples, a rule of thumb is to insert by at least four times the sheath outside diameter. Iron/constantan type thermocouples should show only slightly higher losses than chromel/alumel, but copper/constantan couples show much greater losses due to the high thermal conductivity of copper. Greater losses are also met if close-fitting refractory sleeves over the wires are used. Long pockets reduces the loss; Fig. 5.4 shows an example of using a long pocket in a restricted situation. Long stainless steel pockets are obtainable from thermocouple suppliers [see catalogues, (**9**), (**10**)]. In all couples, conduction loss is reduced if very fine wires are used but there may be drawbacks regarding fragility and possibly heating due to instrument currents as noted above.

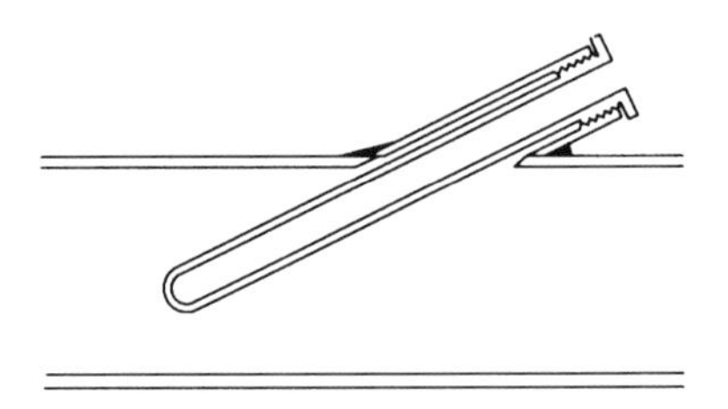

Fig. 5.4 Long pocket for temperature measurement

5.3.6 Radiation and high speeds

At high temperatures there can be radiation losses from the junction, when gases flow in non-lagged ducts. Special couples can be made up with refractory radiation shields. At high gas speeds some consideration should be given to which temperature one requires; the most straightforward to understand is the stagnation or total temperature. To approach this a cup-shaped shield open at the front may be put close to the couple; it may also have a small exit hole downstream. A couple without such a shield could attain something like the adiabatic wall temperature. The distinction as to which temperature is required would be particularly relevant when the efficiency of a machine is deduced entirely from pressure and temperature values, which happens in turbomachinery where direct power measurement would be too costly (this is an instance where temperature data should be as true as possible, for comparison between different test set-ups). For very high gas speeds there is an error due to adiabatic compression of the gas in front of the testing body; this is quoted by BS1041 (7) as being roughly 10°C at 200 m/s for air at atmospheric temperature, and goes up with the square of the speed, though in important cases a fuller thermodynamic analysis should be attempted.

5.3.7 Differential thermocouples

In some cases the difference between two temperatures is more important than the actual level of temperature, chiefly in heat flow measurement. In such cases one instrument and its possible error can be eliminated by using a two-junction set-up as in Fig. 5.5. A form of voltmeter is used, not a meter calibrated directly for temperature, because of the cold junction compensation arrangement mentioned earlier. The approximate temperature levels must be known so that the emf per degree can be checked from tables, allowing for possible non-linearity.

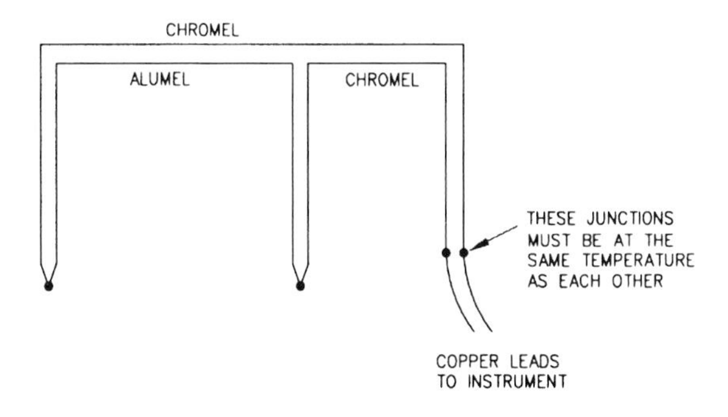

Fig. 5.5 Differential thermocouple using a two junction set-up

5.4 Resistance thermometers

5.4.1 Description

The most usual resistance thermometers use platinum against one of two platinum/rhodium alloys. The metals may be very fine wires wound on a ceramic carrier, secured locally with a refractory cement, or deposited on a non-conducting base as thin films; another form is helical coils loosely held in a multi-bore alumina tube. The most usual cold resistance is 100 ohms at 0 °C. The active bulb is typically 25 mm long, including the filament carrier and immediate protection. Further protection may be a refractory wash fused on, which tends to slow down the response rate. Normally all one sees of the bulb is the stainless steel or Inconel protective outer sleeve. Some versions have small holes to allow circulation as for gas thermocouples, for better thermal contact with the medium being measured.

Platinum resistance thermometers are regarded as a more accurate method than thermocouples; the size of the bulb automatically ensures some extent of averaging, more so than thermocouples which tend to be tip sensitive. For this reason thermocouples are generally more appropriate for localized and surface temperature measurement than platinum resistance thermometers. Thermocouples also tend to be cheaper, more robust, have a wider temperature range and a faster speed of response. However, the platinum resistance thermometers have an almost linear output, better long-term stability and are less susceptible to electrical noise.

5.4.2 Wiring methods

In the laboratory the leads resistance may well be less than 0.1 ohms and undergo little change of temperature. It may either be ignored or allowed for mathematically. In large installations the control console may be far from the sensor and the leads may be at various uncertain temperatures. Even if a transmitter (signal conditioner) is used, this may be placed well away from furnaces, etc. The recommended practice is to arrange for temperature compensation, using the three- or four-wire method.

To understand this one must consider how the resistance is measured. Although disguised in a grey box, the principle is that of the Wheatstone bridge. The description here assumes that the Wheatstone bridge is equivalent to that described in Chapter 7 (Section 7.1.6). Figure 5.6a shows the uncompensated two-wire method, where the resistance thermometer bulb is simply wired in as the unknown resistance. In Fig. 5.6b the three-wire method is shown. The lead L_1 is in series with the bulb, the return lead L_2 is in series with the variable resistor. The bridge **must** be wired 1 to 1 for this purpose. The middle lead goes to the galvanometer. When the bridge is balanced, the resistance sum $X + L_1$ is equal to the resistance sum $V + L_2$ in the other arm. We take care to make the leads L_1 and L_2 identical and run them along the same route. Since $L_1 = L_2$, $X = V$ which is what we require; the leads' resistances are compensated out.

There are two possible four-wire systems; Fig. 5.6c shows the dummy lead method using four identical leads, one pair to the bulb and the other pair going the same way and merely coupled together at the hot end. It seems to have the same compensating effect as the three-wire system; the galvanometer lead is doubled but since it carries zero current at balance, it is difficult to see much improvement in this over method (b); perhaps it is easier to understand. Some instruction brochures show method (d).

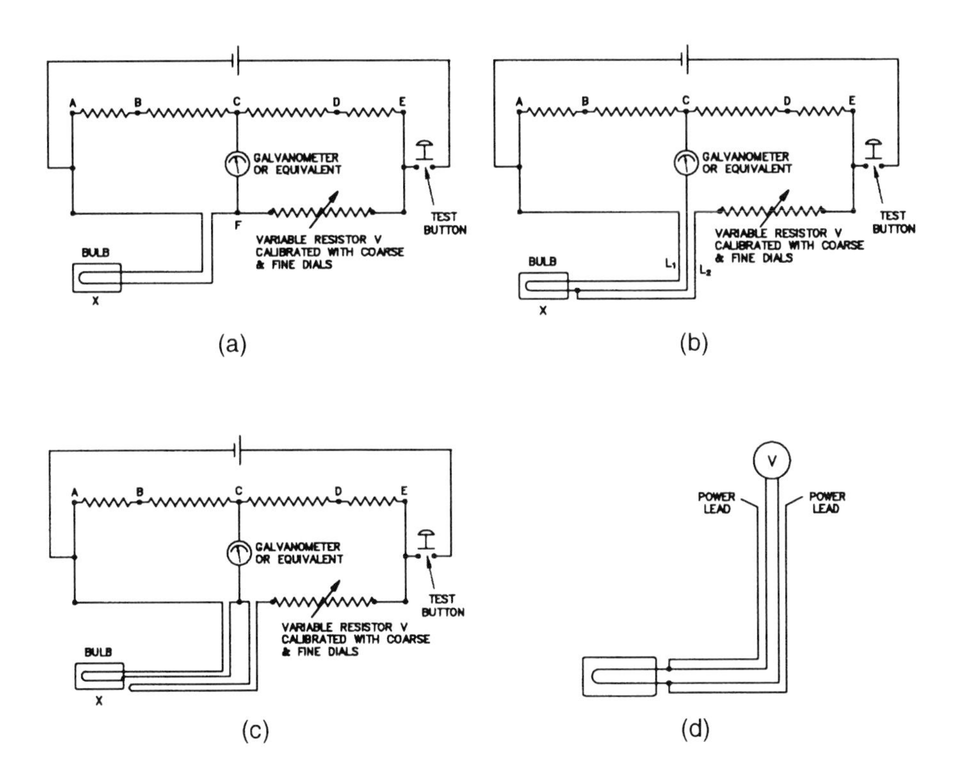

Fig. 5.6 Methods of wiring resistance thermometers

5.4.3 Self-heating

One of the suppliers recommends that the current in the sensor should not exceed 1 mA. Some instruments seem to cause a current of 2 mA in the system; presumably this is split equally between two sides of the bridge. In general, self-heating works in the opposite direction to the conduction loss down the sheathing or down the wires but it does make for uncertainty. In low (sub-ambient) temperature measurement self-heating and conduction errors will be additive rather than opposing. In important cases it may be justified to set up a comparative experiment to estimate the possible magnitude of the effect, as discussed under thermocouples. If the resistance seems to drift upwards this may indicate self-heating.

5.5 Surface temperature, contact, and non-contact methods

Surface thermocouples are obtainable incorporated in a thin strip of a good thermally conducting metal, held on to a surface so that it does not greatly affect the heat flow conditions. The other method is to make the thermocouple very small and insulated on the outside, to avoid conducting heat away. An alternative which has become well known is the thermal imaging camera which translates the temperature of a small scene into a colour picture; the scene can be a face, a body, a machine or electrical equipment developing hot-spots, or the ground when looking for hidden or missing persons. The signal is recorded using conventional video recorders, and can be analysed subsequently to quantify the temperatures on various parts of the image.

Some problems arise when trying to measure the temperature of moving surfaces. Sliding type thermocouples are available, with a sprung element which presses against the moving surface, which works fairly well if a good contact can be maintained. Another possibility is to have the device moving with the surface being measured. The signal then has to be transmitted to the stationary recording device. For reciprocating movements, flexible leads may be feasible, but for continuous rotation, slip rings or radio or infra-red telemetry are required. Slip rings can be troublesome due to the low voltages involved.

In one test rig application it was necessary to measure the temperature of a specimen reciprocating at 30 Hz over a 50 mm stroke, which presented a challenge. Mounting a platinum resistance thermometer (PRT) on the reciprocating head was necessary, and radio transmission of the signal was considered. However, it would then have been also necessary to reciprocate the power source for the transmitter. A better solution was a flexible cable to the recording device, but this failed after a very short time in operation. A higher duty cable in conjunction with a cable guiding device (as used on robots) gave satisfactory results. This technique of a flexible cable with a suitable supporting and guiding linkage has also been used successfully to measure temperatures in the pistons of IC engines. In this hostile environment the life of the cable was finite, but sufficient for the purpose of research and development work.

A non-contacting method of temperature measurement is an infra-red optical method, particularly useful for moving surfaces, very high temperatures, or where a contact method would contaminate the product. The instrument can focus on to quite small spots and interpret the target temperature by analysing the infra-red spectrum. Each instrument can cover a certain band of temperature. Being a non-contact method there may be anomalies of interpretation with surfaces of special reflectivity or emissivity properties such as polished metals. White paint is not necessarily highly reflective in the infra-red. The emissivity can be affected by the temperature, angle of measurement, and geometry of the surface itself. The thickness of the surface and its transmissivity may also have an effect.

On any particular surface type the method may be calibrated by a subsidiary test in which a similar surface is provided, at a temperature known by other means, near to that under investigation, so that the readings can be compared. If the problem is temperature measurement of a moving surface, for example determining disc temperature during a pin-on-disc wear test, then calibration may be done with the surface stationary and a surface temperature probe to measure the temperature, then the emissivity value adjusted to match.

Other methods suggested are to cover part of the body being measured with masking tape (emissivity 0.95) or dull black paint (emissivity 0.98), then measure an area next to it on the body, and adjust the emissivity setting to give the same temperature reading. If paint is being used, then obviously one must ensure that it will withstand the temperature. For suppliers of such non-contact instruments see references (**11**) and (**12**).

An arrangement has been described for measuring engine piston temperatures using the infra-red non-contact method. A hole is cut in the cylinder wall and fitted with a window which transmits infra-red radiation. The infra-red detection instrument views through a stroboscopic disc or an equivalent electronic shutter. By varying the phasing between engine and stroboscope, various parts of the piston can be observed. The method can be verified by a subsidiary test with a static piston at one or more known temperatures but with the detector and stroboscope operating as in the real test. This should cover emissivity problems provided the dummy piston has a similar finish and oil film as the working one. The verification would also cover any effects due to the short exposure and due to the detector viewing the blank part of the stroboscope disc intermittently.

Problems to avoid when using this measurement method are the presence of any intervening absorbing gases such as CO_2 or water vapour, which will significantly lower the readings. Smoke or dirty window glass will have a similar effect.

5.6 Thermistors

Thermistors are small semiconducting devices whose resistance changes with temperature, allowing them to be used for temperature measurement. They have a high sensitivity and small mass leading to a short time constant, but are non-linear and their characteristics vary from sample to sample. One of their main uses is in temperature control, where their non-linearity is not as important. In electric motors they are often situated in the field windings to initiate a shut down in the event of the overheating and prevent damage to the motor.

Thermistors are also used in refrigeration systems, where their resolution, fast reaction time and ability to work at low temperatures and high pressures are paramount. For example an electronic type thermostatic expansion valve operates on feedback control from a thermistor. Condensor fan speed control may be temperature operated, from a thermistor mounted in a thermostatic pocket in the condensor tubing and set to say 40° to maintain the condensing temperature. Here the fast response time of the thermistor is critical, as the compressor pumps in a good deal of heat, and between this and the adiabatic compression the temperature could easily rise 20° in a couple of seconds, and the vapour pressure shoot up accordingly.

5.7 Temperature measurement by surface coating

In one example, in which the specimen to be measured was very inaccessible due to space limitations and the necessity of re-aligning the specimens, coatings of reference materials with known melting points were used on the surface of the specimen. The point at which the coating melted then indicated the temperature. This method was used in electron beam material interactions when trying to separate out the effects of the electron beam induced damage processes (i.e. radiolysis and knock-on damage), from thermal annealing effects. A range of different materials with different melting points were tried in a systematic fashion, and then the results obtained were compared with computed values.

The same principle is used for measuring temperatures in the pistons of IC engines, where small plugs of various low melting point alloys are inserted. This method is described under engine testing, Section 10.6.

Another well-known method is the use of temperature-indicating paints and pigments. These are often in the form of a self-adhesive sticker which changes colour when a certain temperature has been reached or exceeded. One sticker may incorporate several different paint spots or bands corresponding to different temperatures, and from this the actual maximum temperature reached can be determined to within a few degrees.

Similar stickers which work using a thermochromic liquid crystal are available readily and very cheaply. The non-reversing version showing when a certain temperature has been exceeded may be used when despatching sensitive goods, and can be checked on arrival.

Chapter 6

Fluid Flow Measurement

6.1 Open channels, rivers, and streams

The basic weir and vee-notch techniques are well known, and flow rates are calculated by well-established formulae found in hydraulics textbooks or BS 3680 Part 4 (**13**). The main problem lies with establishing the height of the upstream free surface above the notch or weir. In a laboratory set-up this is straightforward, but in the field it may be less so. Another similar technique is the flume, a streamlined flow restrictor, which is more complex to build but which gives less pressure loss.

For rivers, unless a weir already exists, building one to measure the flow may be expensive, or may be unacceptable for navigation or other reasons such as free passage for fish. Various other techniques are available, with varying degrees of accuracy.

Most methods rely on measuring the cross-sectional area of the river at a certain point, and measuring the average speed of water flow through this cross-section. If there is a well-defined cross-section, for example a section of flow contained in a concrete channel, then cross-section measurement is relatively easy. If the flow is in a natural river bed, then measuring the cross-section can be very imprecise.

Measuring the speed of flow can be done by timing the progress of surface floats. Oranges are recommended as they are widely available, expendable, non-polluting, quite visible, and float low in the water and thus respond to the water speed rather than wind speed. The average speed of flow can be calculated approximately from the surface speed. If greater accuracy of speed measurement is required, then a propeller type flowmeter can be dipped into the flow at varying positions and varying depths. This is ideally done from a bridge above the river to avoid disturbing the flow.

There is an alternative, not widely known technique which does not require measurement of the flow cross-section or the flow velocity. This is the 'salt gulp injection' method, which requires only a small quantity of salt solution, a hand-held conductivity meter, and a stopwatch. A measured quantity V_u of salt solution is made up, typically a few litres for a small river, perhaps one litre or less for a stream. The conductivity C_u of this solution is measured using the meter, if necessary diluting a sample by a factor of 10 or 100 with clean water to bring the conductivity within the range of the meter. Next, the conductivity meter is held in the flow, at any position, and the entire quantity of salt solution is thrown in the river at a point sufficiently far upstream to ensure good mixing. Readings of the conductivity C_i are taken at intervals T, say every 5 or 10 seconds, and these will rise to a peak then fall back again to the background level C_o. The flow rate is derived from the simple equation

$$\text{Flow rate, } Q = \frac{V_u \times C_u}{T \times \sum (C_i - C_o)}$$

To check that complete mixing has occurred, one can repeat the process with the meter in a different portion of the flow, or with the salt solution thrown in further upstream. This process is mainly intended for one-off measurements, but there is no reason why it could not be used on a regular basis, or even automated. Although particularly useful for rivers, the method could also be used for pipelines as a low-cost alternative to a flow-meter, or as a periodic cross-check.

Case Study

Continuous measurement of flow in rivers by ultrasound

It was required to monitor continuously the flow rate of the river Wye, for both resource planning when supplying water to some towns downstream, and for flood warnings. One method considered was to build a weir and use the levels upstream and downstream of the weir to calculate the flow rate. However, there was a certain type of fish known as the shad, which had to swim up the river in order to breed and would not have been able to get past the weir. It was therefore decided to use ultrasonic methods for flow monitoring; this consists of setting up ultrasonic transmitters and receivers across the flow with the equipment on one side of the river being downstream of that on the other side. When the ultrasound wave is transmitted from the upstream to the downstream station the sound travels faster than going from downstream to upstream, and the time difference is used to calculate the water flow speed. If the cross-sectional profile of the river is reasonably well characterized, the volumetric flow rate can be deduced.

Unfortunately the shad also had a problem with ultrasound, in that it was extremely sensitive to a certain range of frequencies, and would have found the transmitters as impossible to get past as the weir. Luckily a set of experiments had been conducted to determine which frequencies the fish were sensitive to, and the system was able to operate successfully outside this range without affecting the fish.

There are a number of possible errors associated with this method of monitoring water speed. Some of the main ones are turbulence, which can completely disrupt the signal, and differing temperatures of the water, which causes the ultrasound wave to travel at different speeds. Another is 'skew flow', where the water is not travelling parallel to the river banks; a way of accounting for this is to use four stations, with an upstream and a downstream station on either side of the river, so that the average of the two results will be correct.

6.2 Steady flow in pipes

The meaning of steady flow, in this context, is that the pressure difference used to measure the flow is able to keep pace with any changes in the flow rate. The pressure drop is a function of speed squared, hence a mean steady reading represents a mean value of speed squared, while the average flow depends on mean speed (not squared). A correction would require a knowledge of the fluctuation's wave form which may be sinusoidal or more complicated. Also, inertia in manometer lines may distort the reading.

It can be important to establish what constitutes sufficiently steady flow. If the flow rate is pulsing rapidly, for example the output flow from a single cylinder reciprocating compressor or pump, or the intake air flow to an IC engine, then the monitoring method will not be able to track the waveform of the flow, but will merely register an 'average'. If the monitoring method is by orifice pressure drop, which depends on the square of the flow speed, then one will effectively be measuring RMS (root mean square) flow rather than true average flow. The discrepancy can be quite severe, depending on the magnitude of the pulsations.

The solution for rapidly pulsating flows is either to smooth the flow by means of a large chamber, or to use a monitoring method appropriate for pulsating flows, as described in Section 6.5. In general, if the pulsations are small, say less than 5%, or alternatively they occur on a slow cycle which can be tracked by the monitoring method, then the methods described below for steady flow can be used with reasonable accuracy.

6.2.1 Orifice and similar meters

The chief traditional method uses a constriction in the pipe. BS 1042 (**14**), also ISO 5167-1 (**15**), gives details of constrictions for which reliable data are available so that the flow rate becomes predictable to a high, stated level of confidence. The standard devices, data, calculation procedures, and tolerances have been refined over many years.

The constrictions available comprise orifice plates, nozzles, and venturis. Figure 6.1 shows some of the leading features, the rest are given in the standard. The advantages of using a constriction are (a) a single reading is sufficient rather than a multi-point survey, (b) the designer can choose the size to give a convenient range of pressure drops, and (c) the speeding-up makes the flow profile largely uniform and predictable.

The presence of a constriction obviously demands an increase in the supply pressure, to make the fluid accelerate. The pressure drop between the upstream tapping and the other, in the quiescent downstream space or at the throat of a venturi, is used to calculate the flow rate, as described in the Standards, which give the coefficients of discharge, allowance for approach velocity and for expansion due to pressure change (in gases), further constructional details and tolerances.

Whilst the designer has considerable freedom in choosing the constriction size there is always a compromise between obtaining a large reading and avoiding excessive overall pressure loss. Very small pressure differences call for inclined or differential manometers, or manometers with a microscope to observe the level. The pressure loss is less than the manometer head, since all the devices give some pressure recovery in the downstream pipe. The classical venturi and venturi-nozzle systems give the highest, around 75% of the drop used for the measurement. Orifice diameters greater than 0.8 of the pipe diameter are generally avoided, as the level of accuracy becomes less at such large values, though of course the overall pressure loss is also less. A nozzle or bellmouth used at the

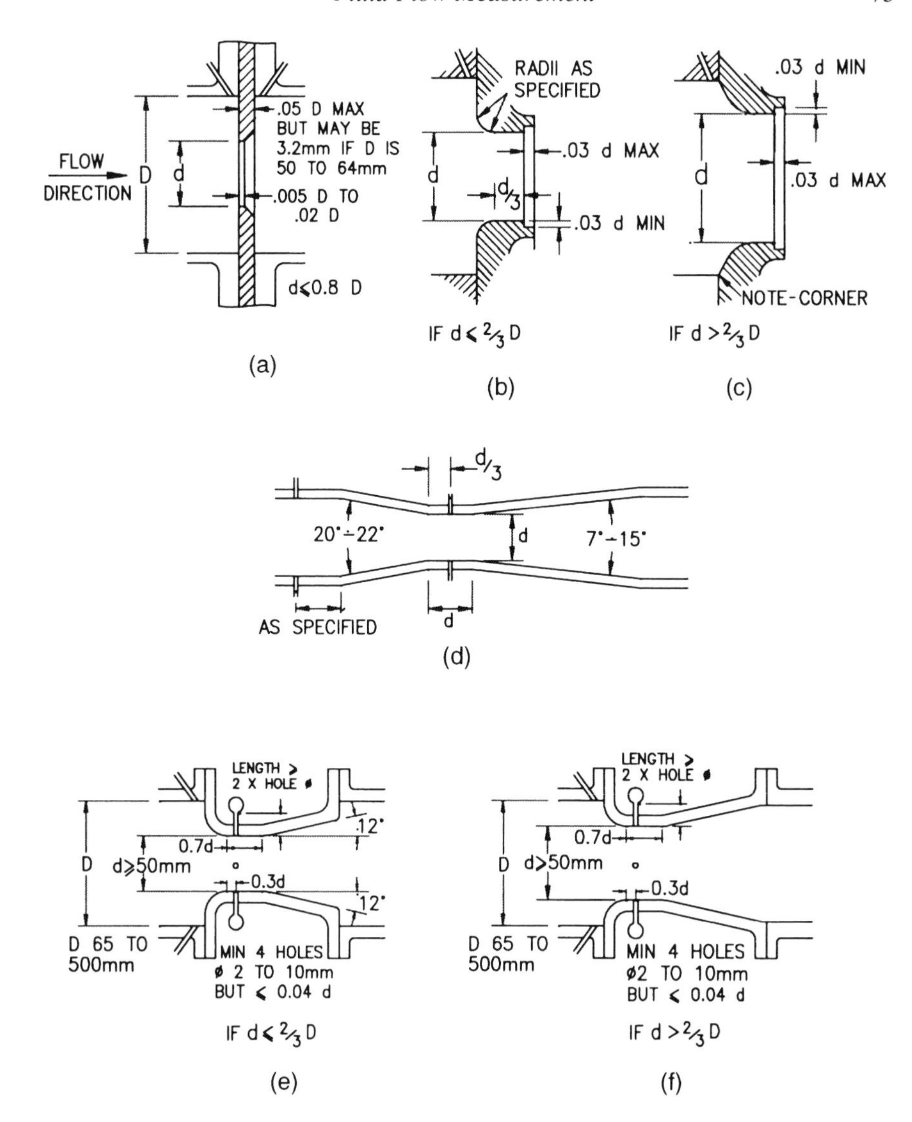

Fig. 6.1 Constrictions in pipes; orifice plates, nozzles and venturis

pipe entry where possible has the advantage of giving a flow measurement by measuring the pressure immediately downstream of the entry, but without further constricting the flow and causing pressure loss.

To obtain a reliable reading a certain calming length must be provided upstream of the constriction, ranging from 10 to 80 pipe diameters depending on the severity of the disturbing feature (entry from free space, diameter change, bend, globe valve, etc.). About half these lengths will suffice if a 0.5 % extra uncertainty is acceptable. As alternatives to these lengths, various acceptable flow-straighteners are described in the standard BS1042.

The readings may also be affected by features downstream of the constriction such as centrifugal fans or pumps which give rise to an upstream pre-swirl. In wind tunnels the type and position of the fan (including whether it is upstream or downstream of the working section) affects the uniformity of the flow. Use of a honeycomb section upstream of the working section can help to straighten it. Curvature of the tunnel wall or sharp features can cause the flow to separate from the wall. Uniform velocity flow parallel to the wind tunnel wall, and with a relatively small boundary layer can be critical to wind tunnel measurements; it is useful to check this by performing a pitot-traverse across the working section.

The simplest constriction is the flat-faced, sharp-edged (also called square-edged) orifice plate, Fig. 6.1a. It is easy to make and easy to remove from the pipe-work for inspection, cleaning, or replacement.

In older installations the pressure tappings may be found at one pipe diameter's distance upstream and half a diameter downstream. The present standard expects the pressures to be taken at the corners, at four or more holes connected together externally by a ring-main (piezometer ring). In low-pressure work this can often be made of flexible plastic tubing and plastic, glass, or metal T-pieces. In high-pressure work, obviously, stronger materials must be used even though the pressure differential used for measurement may be low. When measuring small differentials at high overall pressures it is particularly important to look out for leaks since the frictional pressure drop due to a leakage flow would falsify the reading.

It is advisable to take the tappings at 45° or other angles away from the bottom if the pipe is mainly horizontal, to avoid deposits blocking the holes. Arrangements for draining or venting the pipework are normal industrial practice and are kept separate from the measuring lines. Nevertheless small drain holes are allowed in the orifice plate, otherwise when gas flow is being measured they may fill up with liquid to the orifice edge and give incorrect upstream or downstream speeds; this would invalidate the flow formulae. With liquid flow measurement the converse may take place, a build-up of air or gas. Presumably the hole would be placed at the top, acting as a vent-hole.

An alternative form of pressure take-off can be made by machining the flanges out to create a small gap and an annular plenum chamber, before and after the orifice plate, from each of which a single pipe is taken to the measuring device.

This type of orifice plate is intended for pipes over 50 mm bore. The discharge coefficient ranges from about 0.62 to over 0.8, depending mainly on the ratio of orifice diameter to pipe diameter. For low Reynolds numbers, as with smaller pipes or relatively viscous liquids, the standard gives conical or rounded entry orifice plates (not shown).

Standard nozzles such as shown in Fig. 6.1b and c or the similar long-radius nozzles with elliptically curved entry shapes (not shown) have discharge coefficients very close to 1 and are therefore less subject to uncertainty. The recess in the downstream face appears to be important, presumably to avoid back-flow. Its length and radial width are about $0.03d$ and the width-to-depth ratio must not exceed 1.2. The long-radius design has no recess but calls for a wide, flat end-face.

There has been considerable research into nozzle shapes, both on theoretical grounds and for optimum performance from a given pressure in fire-fighting situations; the nozzles shown here are close to optimum performance but are mainly chosen to facilitate accurate manufacture.

When good pressure recovery is important, the classical venturi form, Fig. 6.1d may be used, or the nozzle–venturi combinations 6.1e and f which are much shorter. The standard calculations apply to these only if the Reynolds number in the pipe is between 1.5×10^5 and 2×10^6. It is also important to pay attention to the rules for the pressure tappings in the throat. There must be not less than four but an all-round slot is not allowed. It is thought that the reason is to avoid risk of backflow from the region where the pressure is recovering as the speed falls. In extreme cases backflow may reduce the coefficient of discharge by an unknown amount; furthermore the designer of the overall system may be relying on good pressure recovery.

There are two sources of error. The first is corrosion or scaling up, in pipes which have been installed for a long time; this applies chiefly in liquids and flue gases. Therefore the installation should be such that a section of pipe containing the device is accessible for inspection and cleaning.

The second error concerns the lines leading to the manometer or other recorder. When liquids are being measured, the main error risk is trapped air or gas bubbles creating a false head. Bubbles can appear in an originally clear line, perhaps due to temperature changes; where feasible the lines should be transparent, inspectable, and laid out avoiding long horizontal runs. Even in a horizontal line, an air bubble occupying the whole line cross-section can give slightly false readings due to uneven surface tension which can make the bubble stick in place and give a slightly false manometer reading. High points should have venting valves.

When measuring gas flows, liquid can condense out and must be allowed to drain away, not occupy the whole line which would give a false pressure drop. Obviously, condensation is prevalent when steam is involved, but it also occurs with compressed air, or gases from most combustion processes such as engines, boilers, or furnaces. Again it is best to avoid long horizontal runs where liquid may lodge, also vertical runs. Sloping lines are best and it may be necessary to use catch-pots with drain valves.

An error can result from thermal expansion of the fluid, for example when the restrictor is in a chimney well above the measuring instrument. It may happen that one line is close to the chimney and gets significantly warmer than the other; then there is a static pressure difference due to the different densities. Figure 6.2 shows a flue fitted with a venturi meter and pipes leading down to an inclined manometer giving a magnification of 5:1, such as may well be used to monitor exhaust gas flow in practice. Assuming the pipe nearest the flue to be 20 m high, and at 120 °C, whilst the other pipe is 20.5 m high and is at 40 °C, we proceed as follows.

If the density of the gas in the pipes is the same as air, for example, which it may well be as it will have less oxygen but more CO_2, then the pressures at the instrument are greater than those at the venturi meter, by different amounts.

Taking the warmer pipe, the density is ($1.29 \times 273/393$) kg/m^3; the pressure p at the bottom exceeds that at the top by ρgh, i.e. density $\times$ g $\times$ height. This works out to 175.8 N/m^2. For the cooler pipe the density is ($1.29 \times 273/313$) kg/m^3. This pipe is 20.5 m high, thus the pressure at the bottom exceeds that at the top by 226.3 N/m^2. This corresponds to a water column height given once more by using $p = \rho gh$ but this time

ρ = 1000 kg/m^3, so at the manometer there is an error of (226.3 − 175.8)/(1000 × 9.81) = 0.00515 m, or near enough 5 mm. As the manometer reading is magnified five-fold by the sloping tube, the reading needs increasing by 25 mm to give the true flow in the flue. The static head due to temperature could produce a fictitious indication of reverse flow.

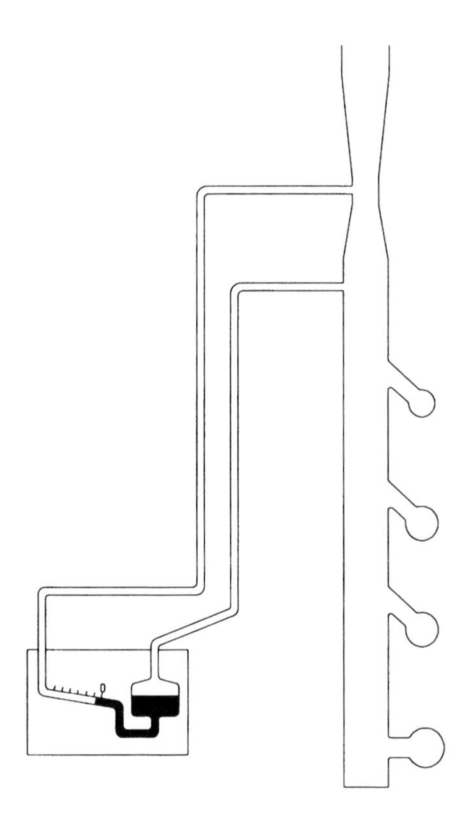

Fig. 6.2 Flue fitted with venturi meter

6.2.2 Turbine meters and rotameters

There are a number of turbine-type instruments using a rotor shaped like a propeller. They are called anemometers if used for air, current meters if intended for water. Some types are for use in the open, some are designed to be installed in a pipe. The speed was traditionally measured by gearing and a counter, but now more commonly by measuring the frequency of passing of the blades. One of the simplest, most robust counting methods is by a small magnet and a reed switch. Others use an interrupted light beam, electrical capacity, or inductance. The light beam can be affected by air bubbles or strong lighting. Some of these meters have a small insert in the inlet pipe to be used at low flows, in order to give a higher jet speed. If it is not inserted the meter may be non-linear at low flow rates. For low flows, remember to re-calibrate the meter with the insert in place.

Sometimes turbine meters for water or other liquids are designed so that the rotating parts are neutrally buoyant. This means that the bearing friction at low flow rates is virtually zero, thus permitting the unit to register very small flow rates.

If a turbine meter is used for a liquid flowing out of a pipe into the open, the pipe may be only partly full. The speed of the turbine may well be that of the liquid but the flow rate is given by speed times cross-section ***of the flow***, not cross-section of the whole pipe. The meter therefore over-reads, perhaps by a factor of 2 if the pipe is half full. This can also occur if the meter is installed at a high point in a pipeline, because an air bubble may collect and cause the meter to run part-full. To avoid both these problems, the meter should be installed on a portion of the pipe sloping upwards in the direction of flow. A transparent port either in or close to the meter would be valuable for checking conditions.

Another error in this situation occurs with intermittent flow in the output from piston pumps. The inertia of the rotor can keep it rotating slightly during no-flow periods, or conversely the rotor speed may lag behind the flow at the start of each stroke. Either way, there will be some error, though not usually very large.

When checking a flowmeter on the output from a reciprocating pump, it should be noted that the flow may be continuous in practice due to the inertia of the liquid, and the ability of the pump valves to permit forward flow even during the return stroke of the piston. On a reciprocating pump driven by a windmill, it was found that at high wind speeds the volumetric efficiency of the pump appeared to be considerably greater than 100%. At first measuring errors of the types mentioned above were suspected, but checks with a bucket and stopwatch confirmed the flow readings. The inertia of the water in the long delivery pipeline was sufficient to keep the water flowing forward during about 50% of the piston return stroke.

For small steady flow rates a Rotameter-type instrument is useful. This consists of a tapered tube, larger at the top, sometimes ribbed internally to guide what is called the float. Figure 6.3 shows the arrangement. The meter is always mounted vertically; the so-called float is only partly buoyant in the liquid (or gas) and is pushed upwards by the flow, up to a level where the gap between float and pipe wall allows the fluid to pass with the pressure drop sufficient to support the net weight of the float. At low flow rates the float rests on a supporting frame which allows the fluid to pass freely. At about 10% of the full range the float begins to rise into the graduated region. When a longer range is required one has to fit two or more meters, connecting whichever is required. They should not be placed in series, as the smaller meter may restrict the flow unduly at high flow rates.

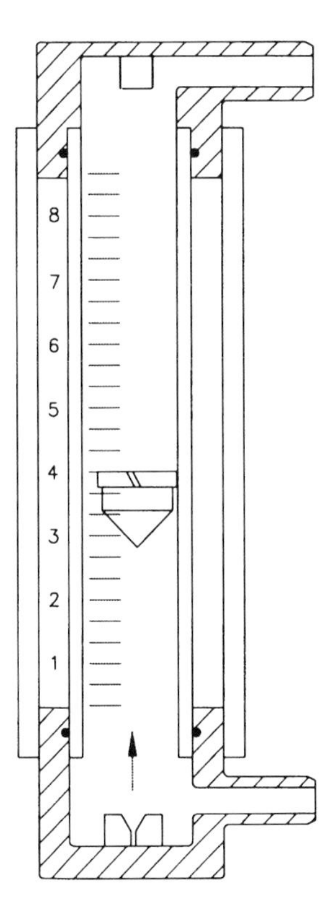

Fig. 6.3 Rotameter type flowmeter

Light-weight hollow floats are used for gases, somewhat heavier floats for liquids. Because of the buoyancy force the calibration applies only to a liquid of a particular density and temperature; for temperature changes a correction factor is supplied whilst for other liquids it is best to order an appropriate float since the calibration also relates slightly to viscosity. The manufacturer will normally supply a conversion chart for use with different liquids. For use with gases and compressible liquids it is also necessary to correct for different pressures.

This form of meter is not suitable for very viscous liquids. The reading can be observed through the glass, but remote reading variants are also available. They are very durable but eventually the edge of the float wears, leaving an enlarged annular area and therefore giving a low reading.

6.3 Positive displacement meters

A favourite mechanism is a C-shaped rotor engaging with a stationary radial vane, in a larger eccentric housing. The flow causes the rotor to orbit and operate a small crank. They are used for domestic and industrial water supplies, or for metering fuel deliveries to or from road tankers. Their chief error is likely to be a slight tendency to by-pass internally and read low, which pleases the paying consumer. On the other hand, being positive displacement they will register every fluid passing through, so if the liquid being measured contains substantial bubbles of air, the meter reading will be higher than the volume of liquid. It has happened that a consumer at some high point in a water main obtained freak high meter readings. This was thought to be due to a build-up of air in the pipeline, perhaps due to pipeline maintenance activities, so that when pressure was restored, a large volume of accumulated air passed into the building's header tank, but registered as if it were water.

Bellows gas meters are also positive displacement devices, with flexible members which actuate a distribution valve as they deflect. They are intended for low pressures and, unless the bellows fail, seem to be highly reliable, potentially reading slightly in favour of the consumer.

Various forms of positive displacement pumps, including gear pumps, rotary vane pumps, and piston pumps are used for measuring or metering flows. They may be acting as metering pumps, i.e. driven by a motor, in which case they will tend to over-read slightly due to internal leakage.

Alternatively they may be pure meters, driven by the flow, in which case they may under-read slightly.

6.4 Non-invasive methods

Several non-invasive methods have appeared using laser or ultrasonic beams passed through the fluid to find the speed, by echoes (Doppler effect) or measuring of transit time. The Doppler methods rely on the presence of targets to reflect the beam, though the natural occurrence of small bubbles or turbulent eddies in the flow may be sufficient in some cases. In other cases, particles or bubbles may have to be deliberately introduced. Accuracy of the Doppler effect systems is of the order of 2%. Doppler effect tends to measure local velocity; thus it depends on whether the flow profile is fully developed, many diameters downstream of the last disturbance.

The transit time methods are somewhat more accurate, 0.5% claimed by some suppliers. However, the liquid must be free from bubbles and solid particles for this method to work effectively.

These methods can be adapted to many situations. For example, laser Doppler anemometry has been used to measure the flow velocity of gases within the combustion chamber of an operating IC engine.

Coriolis flowmeters make use of the coriolis effect on fluid flowing in a curved path. The flow is passed through a U-shaped (or sometimes inverted Ω-shaped) tube, which is deliberately vibrated laterally at its natural frequency. The coriolis force induced by this vibratory movement on the liquid flowing in the curved part of the U-tube causes a phase change between the vibrations of the two legs of the U, and this is detected by sensors on each leg. This phase change is proportional to the mass flow rate, not the volume flow rate, so no corrections are required for density changes. The pressure drop with this device is minimal, corresponding to the viscous losses along the length of pipe only.

Coriolis meters are claimed to cope with fluids containing solid particles, small gas bubbles, etc., and are relatively insensitive to deposits, etc. forming on the inside of the U-tube. They may be affected by large bubbles or gas pockets which could modify the motion of the fluid, so it is important to arrange the orientation to avoid these. The accuracy is very good, claimed at 0.15%, which is comparable with positive displacement flowmeters. As there are no moving parts, the unit is presumably resistant to shock and vibration, except possibly to external vibrations or flow and pressure pulsations close to the natural frequency of the U-tube.

6.5 Pulsating flows

6.5.1 ISO Technical Report TR 3313, 1992 (BS 1042 1.6, 1993)

This report (**16**) defines the distinction between steady and pulsating flow below which time-averaged steady-flow rules may be applied. The limit is when the RMS fluctuation of flow is 5% of the time average. It follows that the limit of fluctuation in terms of the measured pressure difference is 10%. The report discusses the difficulties of establishing the magnitude of these fluctuations; mention is made of compact pressure-difference transducers, with circuits giving RMS outputs regardless of actual wave form. Obviously the feed lines to such transducers must be of sufficiently low impedance and inertia. Expressions are given for processing the results where the fluctuations are above the threshold given above.

The report also mentions direct flow measurement by means of linearized hot wire anemometers, again with circuits giving as far as possible true RMS outputs, not assuming sinusoidal wave forms, etc.

A number of relevant researches are summarized and methods of reducing fluctuations are outlined.

6.5.2 Storage methods

The most definite method for measuring pulsating flows over a period is the total displacement method. For liquids this may be a timed fall of level in a tank, or collection into a small tank over a measured period. When using a large tank there can be an error in the tank's cross-sectional area due to scale or other deposits, or in flat-sided tanks a systematic disguised error since the sides will bulge out more when the tank is fuller, gradually relaxing inwards as the level drops. Clearly, this can give an error in the volume if calculated by nominal cross-section times level change.

One absolute method seen in some laboratories is a gas-holder, a deep, liquid-filled vessel with a rising dome, suitably guided and counterbalanced. This would be used for air, mainly to calibrate other meters. There is a relatively cumbersome form which could be regarded as the rotary equivalent of this, using a multi-chambered rotor half-immersed in liquid. The gas or air is always confined over liquid but the chambers overlap so as to deliver a continuous supply of a fixed volume per revolution.

If a large compressed air reservoir is available, this can be used to supply a quantity measured by the fall in pressure. The method requires three precautions if accuracy is required. Firstly, during the discharge the temperature of the air remaining in the reservoir tends to fall; the temperature must be measured and the volume–mass relation corrected. Secondly, the air being delivered also changes in temperature, and thirdly the pressure gauge should be checked .

6.5.3 Viscous flow meter

Pulsating air flow can be measured using a viscous flow meter, an example of which is shown in Fig. 6.4. The air is drawn through a metal element at low speed, for example a corrugated, spirally wound strip which creates passages of around 1 mm width and perhaps 100 mm length. The length-to-width ratio ensures that the flow is essentially viscous rather than turbulent; hence the pressure drop varies approximately linearly with speed and the average pressure drop closely represents the average speed if the fluctuations are moderate.

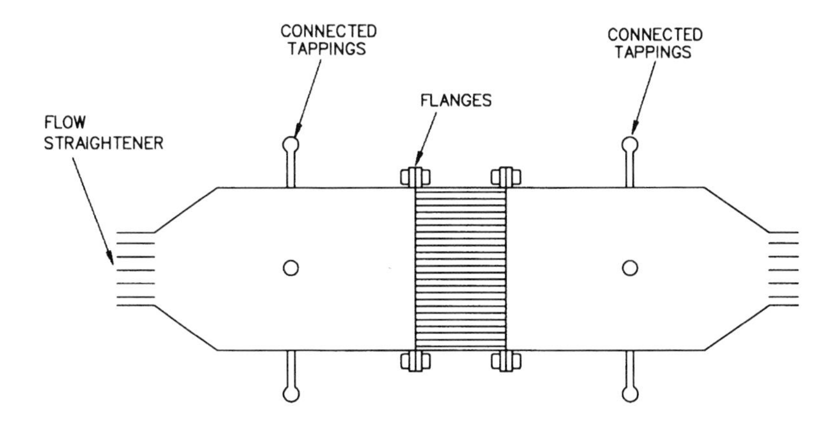

Fig. 6.4 Viscous air-flow meter

If there is a stop–go type of flow (or even partially reversing flow, which does happen under some circumstances) this is not a safe assumption, particularly as the manometer or other pressure-measuring device may not follow the fluctuations correctly; hence severe fluctuations should be cushioned out by having a reservoir in series with the meter. Experiments with a typical viscous flow meter on a single cylinder engine have demonstrated that the pulsations cause an error of 2 to 3%, due to the slight non-linearity in the response of the meter.

The intake needs protecting from dust and draughts. From time to time the stainless steel element can be washed out to remove condensed vapour or other deposits. The approaching air flow should be straight, free from swirl or bends. Serious discrepancies were found when a bend was placed just upstream of the meter.

This type of meter is used extensively in engine testing, but it is now recognized that it can interfere significantly with the induction air flow characteristics, so other methods are now sometimes used, as discussed in Chapter 10.

6.6 Velocity measurement

6.6.1 Short notes on anemometers

Several methods are available, including a three- or four-cup anemometer as used on weather stations, propeller-type anemometers, also deflecting vane instruments and hot wire anemometers. The cup method is omni-directional in the horizontal plane, the cup speed being the net result of the large drag on the cup facing into the wind, opposed by the smaller drag of the other cups. If the meter is partly sheltered so that a cup going up-wind at the time is not meeting the normal wind speed, then the whole rotor may go faster than under the calibrating condition. This means it can actually give an excessively high wind speed reading.

Propeller anemometers must face into the air stream, though small misalignments do not give large errors. Vane anemometers are less accurate since the casing affects the air flow.

Hot wire anemometers depend on the heat transfer from a heated filament, the temperature being deduced from the change of electrical resistance. The supply may be a constant current or a constant voltage; the instrument is calibrated against a velocity found by more fundamental means. Early types were very sensitive to loss of accuracy presumably due to atmospheric contamination; later types using high temperatures are believed to be much more satisfactory. They give a very localized reading which means they can be used for boundary layer studies; for total flow a traverse would be needed.

Thermistors can be used instead of hot wires for measuring very low flow rates. Early electronic rate of climb indicators for gliders used this method to measure the very small rate of air flowing out of a flask as the aircraft gained altitude. By having two thermistors in tandem, the difference in signal between them indicated the flow direction as well as rate.

6.6.2 Pitot-static tubes

These tubes are double-skinned; the centre picks up the total or stagnation pressure of the approaching air flow whilst the static pressure is collected by small holes leading into the annular space. The two spaces are connected to a manometer or some other gauge (electrical gauges are obtainable), to give the pressure difference. Pitot-static tubes if made to a standard design shown in BS 1042 Part 2 (**14**) do not need individual calibration; the values in the standard may be used directly. Figure 6.5 shows one of several acceptable standard designs. The standard includes correction factors for high pressure differentials such as occur with high speeds; with speeds met in the laboratory these corrections will rarely exceed 2 %. The reading is localized and may call for traverses to be taken to give a total flow.

The calculation is only valid if the flow is substantially straight and the tube is well aligned with the flow. The standard gives correction factors for small misalignments. It has been found that convergent or divergent flow tends to affect the static pressure at the tube compared with the wall static in the duct at the same station in the flow. In theory the static pressure should be measured at the same cross-section as the total (pitot) pressure. Thus in such ducts the static pressure at the side-holes in the pitot-static tube is not quite what the theory requires; this may give rise to a small error in calculating the speed. More significant errors result from curved flow. The standard requires swirl (helical motion) to be less than 3°. This may well be difficult to ensure in open flows, especially near aircraft wings or large obstacles. Even the turbulence from a glider tow rope can badly affect the air-speed reading taken from a pitot tube on the nose of the glider. Some quite astonishing readings have been obtained near a Delta-type model wing at high incidence. This wing type is said to create particularly strong vortices.

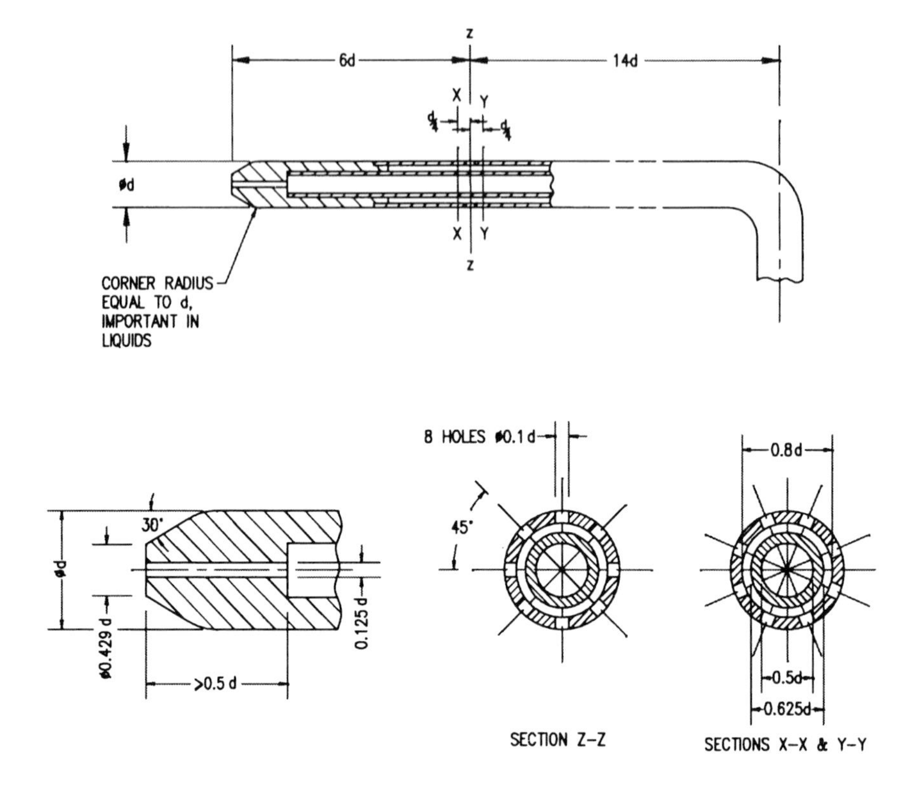

Fig. 6.5 Details of Cetiat pitot-static tube (adapted from BS 1042 Section 2.1 1983)

A pitot tube has been used in reverse, giving a suction instead of an increased pressure. An open-ended tube facing downstream, with a small air bleed and connected to a manometer of some kind was tried out as a water speedometer over 100 years ago on Queen Victoria's yacht. In air this form, the Brunswick tube (or 'total energy tube'), is used in gliders for air-speed compensation purposes on the rate-of-climb instrument. The reverse pitot principle has some different properties from the forward-facing type; it is known that the rear face of a blunt object has a very uneven pressure pattern. The wake is much influenced by any flow from the object; such a flow is called base bleed and can steady the wake, reducing base-drag.

6.6.3 Photographic measurement

If it is feasible to introduce small light particles into the air stream it is possible to obtain direction and velocity (in one plane) at the same time. Soap bubbles (using a reinforcing agent such as glycerine) can be used at low speeds. Small pieces of light plastic foam, chopped downy feathers, thin tissue paper or cotton-wool follow an air flow quite well. The rate of settlement under gravity can be estimated in a separate test. Only a few particles should be introduced per picture, say by blowing in through a small tube. Some experimentation is required to get good results. Thought should be given to the means of collecting the particles after use if the experiment is not in a contained space.

The method is also suitable for water. Suitable particles are small solid (not foam) polystyrene grains, but these are difficult to remove from the apparatus later. Some very good views have been produced using gas bubbles. Strings of small bubbles can be produced electrolytically by short pulses of current in a wire suitably pre-placed. One memorable film using this method has shown how a boundary layer slowly thickens and finally curls up and flows away gracefully downstream.

To produce bubbles enamelled copper wire may be used since the sources of the bubbles can be selected by filing off the enamel locally. Stainless steel, with plastic sleeving, may also be suitable. When there is doubt about forward or reverse flow, e.g. in eddies after some obstacle, some workers have used a modified flash which tails off slowly leaving a comet tail at the end of each image. Such light may also be produced by rotating disc devices in front of a steady light, with tapered slots in disc or stator to make the cut-off gradual.

When measuring the speed of particles from a photograph taken with a focal plane shutter it is essential to take account of the movement of the particle during the shutter sweep time. The exposure time is controlled by the slot width; while the sweep speed is fixed, typically at 2 or 3 metres per second. This is the speed at the shutter; the equivalent speed at the plane of the objects being photographed may be several times higher.

In the 1900s there was a widely published photograph (Fig. 6.6) of a racing car taken with a vertically moving shutter which made it appear to lean forward at about 10°. To further clarify this effect the authors have set up a rotating disc with two diametrically opposed white spots, Fig. 6.7. When rotated the image would give two equal streaks when photographed with an iris-type 'instant' shutter. However, when photographed at 1500 rpm with a focal plane shutter travelling horizontally one streak was greatly lengthened as the shutter partly followed the image, the other being correspondingly shortened.

Typical sweep times of small focal plane shutters are fixed by the mechanism at some 10 to 16 ms. The nominal exposure times set on the dial are achieved by varying the slot width. Not all the SLR cameras use focal plane shutters; some use a window mechanism coupled to an 'instant' (iris-type) shutter.

Fig. 6.6 Racing car photographed with a vertically moving shutter

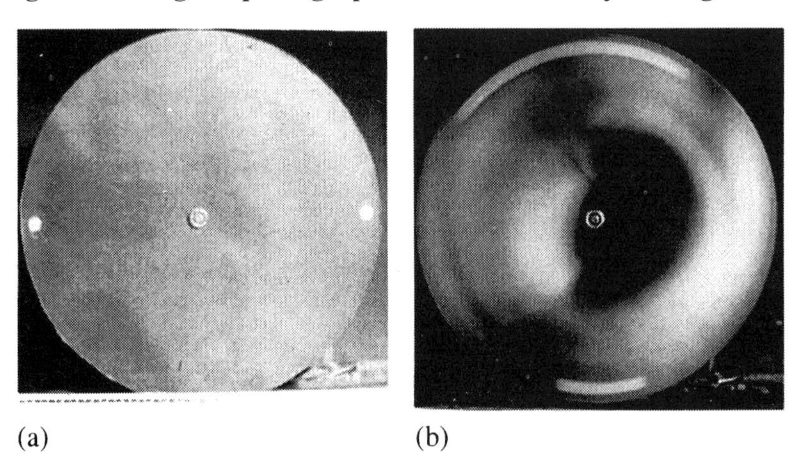

(a) (b)

Fig. 6.7 Rotating disc photographed with focal plane shutter (a) Stationary; and (b) Rotating

Streak pictures are used for direction and speed measurements in fluid studies and could be grossly misleading, particularly in high-speed work, if this source of error is not noted and compensated for.

VHS video cameras are apparently not subject to the same problem; this has been demonstrated with the rotating disc described above.

In aerial photography the object is generally static but the camera is moving. For maximum definition, particularly with high-speed aircraft, it is necessary to compensate for this movement. One system mounts the camera pointing forwards, in an aerodynamic pod, with a mirror at 45° placed in front to photograph the ground. The mirror is rocked slightly during the exposure, to compensate exactly for the forward motion of the aircraft.

In a similar vein, when a still view of a rotating object is required, inverting prisms can be used; in optics these are sometimes called half-speed prisms, as by rotating the prism at half the speed of the object, the image appears stationary. The prism has to be mounted in a rotating carrier with the optical axis (the undeflected ray) on the axis of rotation. Another most important aspect is to use parallel light, obtained from a field lens focused on the object.

These prisms have been used for observing flow in the inlet to turbo-compressors and inside centrifugal pumps with transparent housings. Alignment of the axes of rotation of the prism and object to be viewed has to be quite precise, otherwise wobble of the image occurs.

6.6.4 Special problems with water flow

In a closed duct a pitot-static tube may be used, particularly to find speed variations from place to place, but means must be found to take care of errors due to air in the manometer lines. Feeding both lines with a slow air supply produced enormous disturbances where the air entered the water from the instrument, particularly at the static holes.

In open hydraulic models floats can be used to show direction and speed. They may be surface floats or rather deeper floats with little keels. These were used to confirm a systematic error in a calibration device intended for propeller-type meters, as described in Chapter 2, Section 2.5.5.

6.7 Viscosity measurement

Fundamentally, viscosity should be measured by establishing a uniform velocity gradient and measuring the required force per unit area (shear stress). This requires an accurate and sensitive device and is prone to minor errors at the edges of the moving member.

For industrial purposes there are several forms which use the time of efflux of a certain quantity of fluid from a funnel through a capillary pipe. Provision can be made to run these at various temperatures. There are also certain glass tube instruments shown in BS 188 (**17**) which also use time of flow of a given quantity in a capillary pipe but are enclosed so that there is no evaporation or operator exposure. They are also suitable for immersing in a bath of known temperature.

For high viscosities the capillary pipe method has been used with an applied pressure. There have long been some inconsistencies in results obtained thus and it has been found by two independent investigations (**18, 19**) that a hot slip zone forms near the pipe wall. This might have been expected with oils, since the shear stress here is higher, so the temperature rises preferentially, causing the shear rate (velocity gradient) to rise. Hence ever more energy is dissipated near the wall. Thus the test is taking place (a) at a false temperature and (b) under shear conditions differing from normal iso-viscous theory. This may also explain why designers of oil-hydraulic equipment sometimes find that pipe pressure losses tend to be lower than predicted.

An important property of oil is viscosity increase at high pressures, which is relevant in lubricated contacts (elastohydrodynamics). Ways of measuring high-pressure viscosity include ***low-speed*** flow in capillary pipes with high back-pressure or the Stokes' Law method, measuring the time for a steel ball to fall a certain distance in a pressurized chamber.

Case Study

An oil and grease viscometer

A combined oil and grease viscometer (Fig. 6.8) was designed, to measure viscosities over the range 25–1000 cP for oil, and 1000–100 000 cP for grease. The range was to be covered by two alternative rotors.

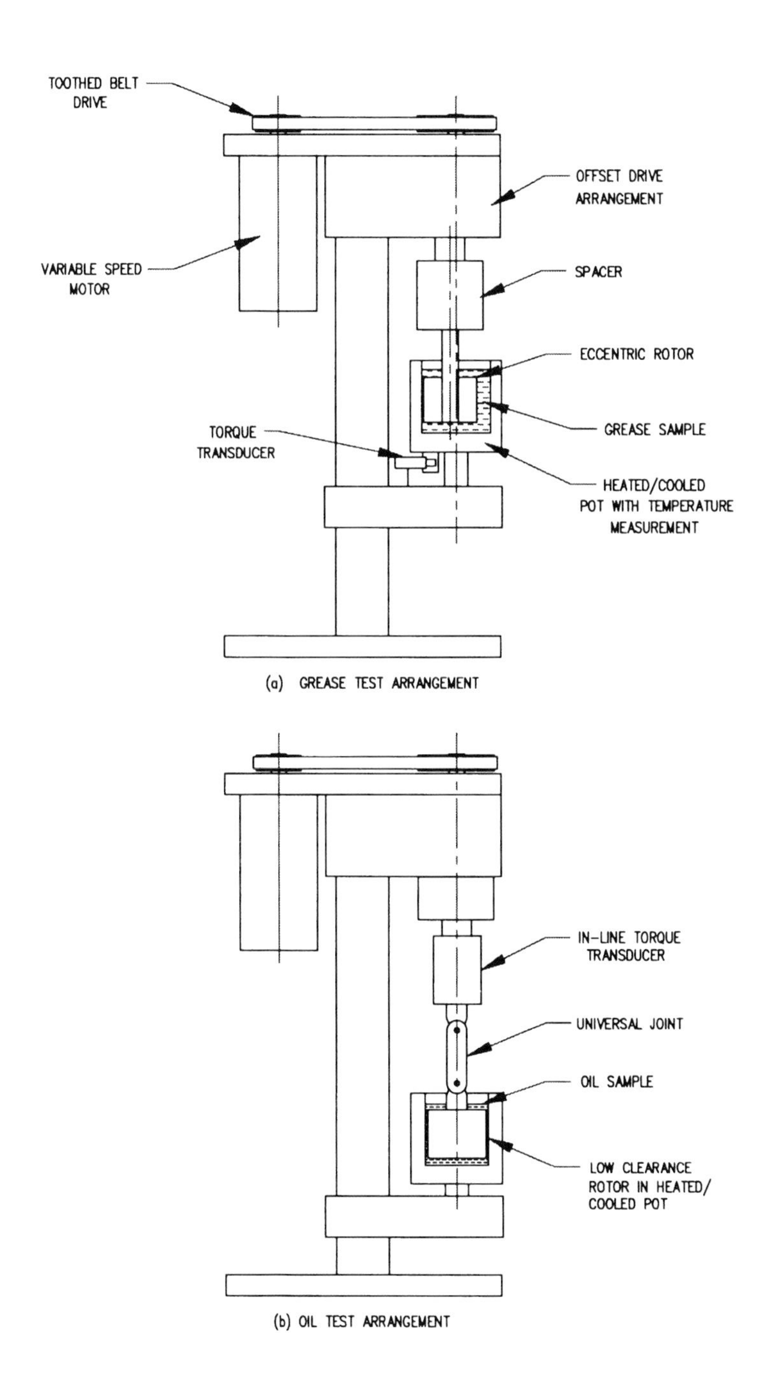

Fig. 6.8 Combined oil and grease viscometer

The requirements for grease measurement were very different from those for oil. The grease had to be subjected to a defined shearing process and the torque measured. The concept was to rotate a bob inside a stationary pot containing a known quantity of the lubricant. However, if a concentric rotor was simply spun in a pot of grease, the grease would tend to 'lock up' i.e. stick to the rotor and rotate with it. The shear between rotor and pot would take place somewhere uncertain in the clearance space. As shown in the figure, an offset rotor was being used to 'work' the grease. The eccentricity imposed a pumping action to the grease to keep it moving consistently.

Previously a standard grease test might consist of working the grease with a piston-type device, then scooping it into a penetrometer and dropping a cone into it to see how far it would penetrate. This has the disadvantages of a less definable working process, and no on-line measurement of the state of the grease, which would be affected by the temperature and would be further worked by transferring it to the penetrometer. Hence there was an obvious scope for a wide-range viscometer combining continuous torque measurement, variable speed drive, and an external heating and cooling facility within one instrument. With the eccentric shaft fitted the shearing and pumping action was simultaneous and under temperature and speed control, and therefore free from the disadvantages mentioned above.

The grease pot was trunnion-mounted and torque measurement was carried out with a load beam; this did not present a problem as the torques were reasonably high. However, the measurement of oil viscosity presented a great deal more difficulty. The minimum viscosities to be measured were very low, so a very small clearance had to be used between the rotor and the pot containing the oil sample. As the oil rotor was interchangeable with the offset grease rotor arrangement it was not necessarily exactly concentric with the pot. It was therefore suspended from a universal joint to allow it to self centre during rotation.

Upstream of the universal joint was an in-line torque transducer, giving an almost infinite resolution on torque measurement. The torque readings at first seemed very high. More investigation showed that a high level of smoothing had been added to the torque transducer signal, and when this was removed and the output connected to an oscilloscope, a cyclic signal in time with the rotation was discovered. This was due to a single universal joint being used instead of a double one, so that the rotor could not fully self centre, and when this was corrected the cyclic signal disappeared.

The readings still appeared too high. The torque transducer reading was zeroed with the rotor inserted in the oil, as it was thought that the buoyancy of the rotor might affect the reading at low viscosities. However, this was found to have little effect; a much larger effect observed was the difference in torque readings at different speeds when rotating in air, due to variable friction in the slip rings. It was therefore necessary to zero the reading with the rotor rotating at the required speed in free air; correcting for this gave consistent results using calibration oils of a known range of viscosities.

For the reasons mentioned earlier; ill-defined temperature and shear conditions across the gap, it is necessary to calibrate these types of viscometers at the speed used for testing and with calibration oils close in viscosity to those being measured. With grease, the viscometer can only be used for comparative measurements with the same instrument, as the results are not directly comparable with traditional, less controllable methods for grease viscosity measurement.

Chapter 7

Electrical Measurements and Instrumentation

This chapter concerns various forms of electrical measurements. Section 7.1 refers to measuring parameters such as currents, voltages, power, frequency, and electrical resistance using meters, and reviews the instruments available. Section 7.2 is involved with the electrical output from transducers, the transducer static and dynamic characteristics, and what happens to the signal downstream, in terms of signal conditioning and processing, and how to avoid introducing errors at this point.

7.1 Review of electrical instruments, mainly meters

Electrical instruments include both portable instruments and fixed panel meters. These can be divided into analogue and digital meters. With the advent of the micro-chip, many meters nowadays use digital processing and electronic circuits to give a digital display of the measurand. However, there is still a place for the conventional moving coil meter, which can be found in many control rooms and on portable test instruments.

The majority of meters interfere with the circuit under test in some way, often by drawing some current, which can be significant. However, with high-impedance meters the effect is usually negligible.

Although the following instruments are well known, the reader may welcome a reminder. This is a summary of the main electrical instruments, including elementary models. Most of these may be found in the catalogues given in references (**9**) and (**10**).

Electric meters indicate voltage, current, wattage (with reference to reactive circuits where watts is not merely volts times amperes), resistance at low voltage, resistance to high voltages for insulation testing, etc. The simplest form takes some current from the circuit under test by imposing an extra resistance. Gross errors experienced tend to be either total failure or wild fluctuations, and the units are susceptible to damage if dropped. Minor errors include parallax errors, stiction (static friction) and non-linearity of scale. The mechanism can serve in a voltmeter or in an ammeter (see below) and in many cases both these and further functions are combined in a multi-meter, the required property and range being selected by a suitable switch. Some advantages are that the effect on the circuit under measurement is highly predictable, at least in the d.c. version, and that no separate power supply is involved except when measuring resistance.

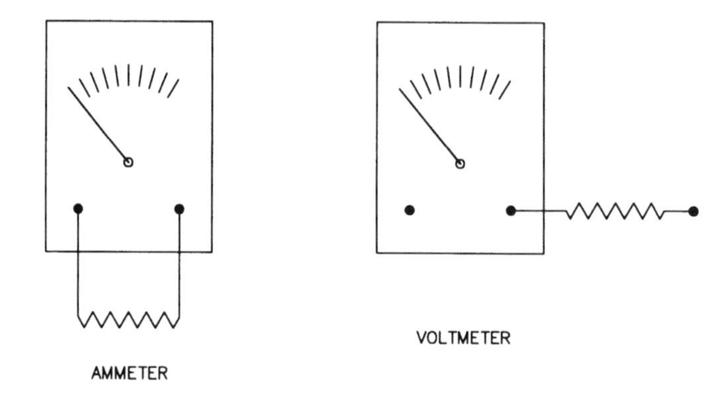

Fig. 7.1 Standard meters with shunts

For alternating current, most of these meters have a rectifier built in, and show a separate a.c. scale. The effect of the rectifier tends to produce an output relating to the average value of current or voltage whilst the scale is graduated in RMS (root mean square) terms which the user normally requires. If the wave-form is non-sinusoidal the scale indication can be untrue. The more expensive multimeters contain microprocessors which examine the wave form, display it on a graphic screen and claim to produce the true RMS value (scope-meter or wave-meter). As noted above, a meter has some effect on the item being observed, although the effects in practice are small and can be allowed for by calculation. The meter movement itself takes only a small current and has only a moderate voltage drop; in ammeters the required ranges are obtained by shunts (accurately known low resistances) which by-pass the extra current round the meter movement. Voltmeters obtain the required range by adding accurate resistors in series. These shunts or resistors are preferably built-in during manufacture to ensure secure contacts, and standard meters with a variety of shunts are available, though for experimental purposes temporary connections are quite satisfactory (Fig. 7.1). The instrument's parameters are normally stated on the dial, the terminals, or in the instructions. Thus any required shunts or added resistances are readily calculable, also the effect of the meter current or voltage drop. In many cases the effect is negligibly small.

Ammeters are normally built with a good capacity for momentary overloads, several times full-scale current, although there is a trade-off between sensitivity and ruggedness. Capacity against over-voltage is rather less since the light, delicate windings cannot be provided with thick insulation, so fuses are used for protection. Some are protected by Zener diodes which cause excess current to spill over through some other route, by-passing the delicate windings; some are further protected mechanically by contacts which trip out if the pointer hits the end stops violently. Most meters are provided with damping against vibration or sudden deflections, e.g. by a liquid-filled chamber around part of the moving member, or by a short-circuited coil in the magnetic field, or by an air vane.

Clip-on ammeters are specially useful for testing imbalance between phases when separate wires are accessible. They are very convenient and have safety benefits for high currents in that they avoid breaking the circuit being measured, but are not normally used below about half an amp. The majority work on flux changes (like a transformer), but some work on the basis of the Hall effect.

Accuracy requirements and general specifications for the various kinds of analogue electrical meters are defined in BS 89 and recording instruments are covered by BS 90 (**20**, **21**).

7.1.1 Moving iron galvanometers

These are the simplest, cheapest, and most rugged form of analogue meter in common use. The field winding is stationary, not needing particularly fine wires. Figure 7.2 shows schematically how it works; the moving part is a light member of soft iron carried in two jewelled bearings like in a mechanical watch. The iron member conveniently forms a counterweight for the pointer. Moving iron meters are usually suitable for alternating current at mains frequency or thereabouts since the magnetic attraction on the iron is equal for forward and for reverse current. Their main advantage is that they can be used for both a.c. and d.c. measurements, and their main disadvantage is oscillations, which need damping. Catalogues list moving iron meters with ranges of 0–1 A upwards or 0–15 V upwards. The moving iron method can be more sensitive than this and has been used for low-cost pocket multi-meters.

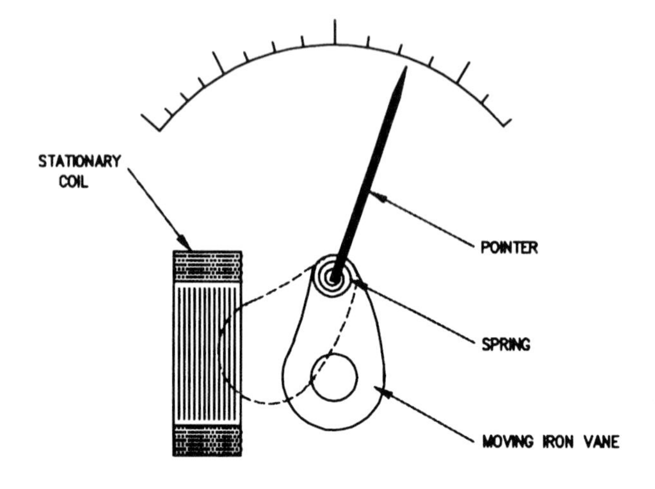

a) ATTRACTION TYPE

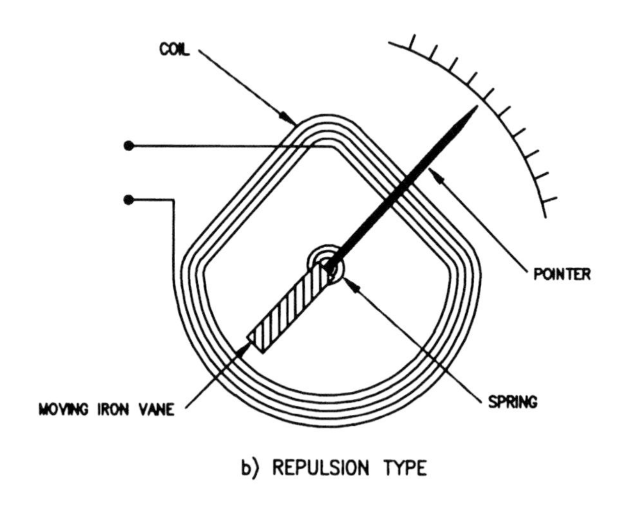

b) REPULSION TYPE

Fig. 7.2 Operation of a moving iron galvanometer

7.1.2 Moving magnet meters

Highly sensitive lamp-and-scale galvanometers are often found in physics laboratories, where a horizontal bar magnet carrying a mirror is suspended by a quartz filament between two coils of wire. At rest the magnet is at right angles to the coil axis. A current in the coils deflects the magnet in either direction, according to the current direction, and the deflection of the mirror causes a focused light beam to project a spot with a dark reference line on to the measuring scale.

A variant of the idea uses a magnet on the pointer spindle for a very simple, tough meter used in battery chargers and older motor vehicles, indicating whether charge or discharge is taking place. They have no windings; the current passes straight through a flat copper bar in the instrument, providing enough magnetic field around it to deflect the magnet. This is like the well-known school demonstration of the way a current affects a compass needle. The zero position is simply determined by a U-shaped pressing of mild steel strip placed around the moving magnet with the pointer at zero. There is no need for a spring restraint, the magnet itself provides the force opposing the field from the current and also holds the magnet and spindle in place axially.

7.1.3 Moving coil meters

Most galvanometers are of the moving coil type, more sensitive, and more accurate than moving iron meters, at only two or three times the cost. The coil can be made such that it gives full-scale deflection with only 50 or 100 μA. A typical moving coil galvanometer may give a full-scale deflection for a current of 1 mA and have a resistance of the order of 300 ohms.

The movement consists of a number of turns of wire on a lightweight frame, in a magnetic field, with a watch-type jewel bearing and hair-spring suspension or a taut-band arrangement which acts as carrier, spring restraint, and current feed. The current causes the coil and needle to deflect. For some purposes central zero is preferred; this is easily arranged at the design stage. The principle is the same, since the action of the fixed magnetic field on the windings acts in either direction, according to the direction of the current.

This is essentially a d.c. method; for a.c. a rectifier or an indirect method (e.g. thermal) is required. A zero adjustment is commonly provided. The meters are obtainable either for building into equipment (panel meters) or in bench-top form for convenience in the laboratory.

7.1.4 Wattmeters

A wattmeter takes account of the phase difference between current and voltage. Such a difference is inherent in induction motors though it can be overcome by some methods. The traditional form is constructed like a moving coil galvanometer but the field yoke is laminated like a transformer. The field winding and the moving coil winding are fed separately; one carries the voltage across the terminals, reduced by a series resistor if required. The other winding carries a current proportional to the current flowing to the load; this may be arranged by a shunt, or since a.c. is involved it may be fed by a current transformer.

7.1.5 Electrostatic voltmeters

These are likely to be seen in a physics laboratory; they tend to be used from 100 V upwards and have the specific property that they take zero current from the load. This may be important in some circumstances. The function of these meters is shown in Fig. 7.3; the static charge exerts an attraction on the moving vane, thereby deflecting the pointer.

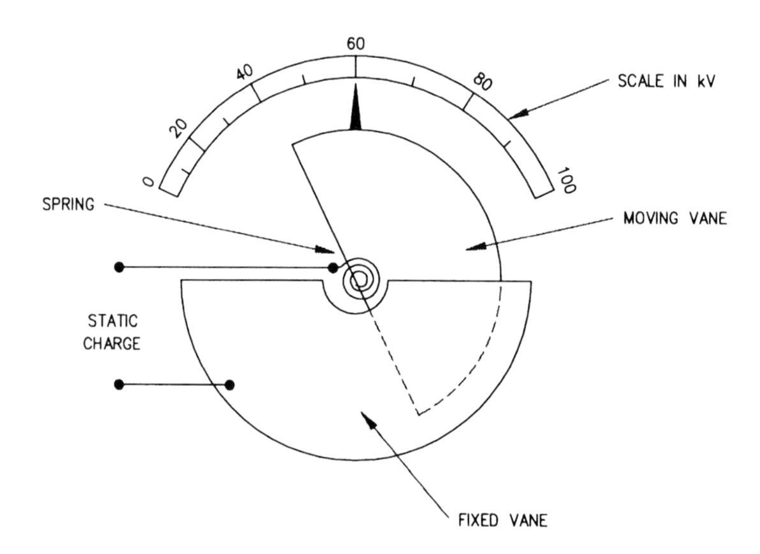

Fig. 7.3 Operation of an electrostatic voltmeter

7.1.6 The Wheatstone bridge

The principle of the Wheatstone bridge for measuring resistance is shown in Fig. 7.4. The unknown resistance is shown at X, whilst at K a known resistance must be placed. A uniform wire is connected between A and C, over a graduated scale. From the junction between X and K a wire is taken to a sensitive centre-zero current detector, possibly of the lamp-and-scale type mentioned above.

From this, a flexible lead is taken to a hand-held probe B which may be made to contact the wire anywhere. The whole is connected to a battery as shown. This simple apparatus may often be found in school laboratories. When the test button is pressed momentarily, the detector deflects in either direction. By trial and error a point on the wire is found which causes no deflection when the button is pressed. Then the bridge is in balance and the unknown resistance can be calculated, since X/K = AB/BC (meaning the lengths of wire between the points A–B and B–C). The virtues of the method are that no persistent current flows, so that there is little risk of temperature rise to alter the resistances.

A more convenient form is shown, somewhat simplified, in Fig. 7.5. The wire is replaced by fixed accurate resistances. The known resistance takes the form of an accurately calibrated variable resistance. The action is described on the diagram itself.

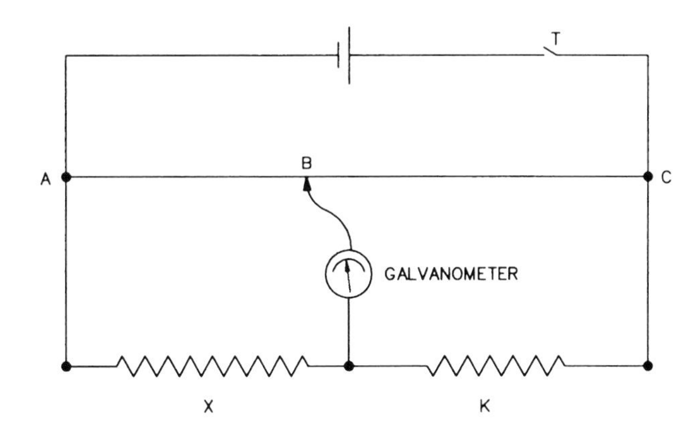

Fig. 7.4 Using a Wheatstone bridge to measure resistance

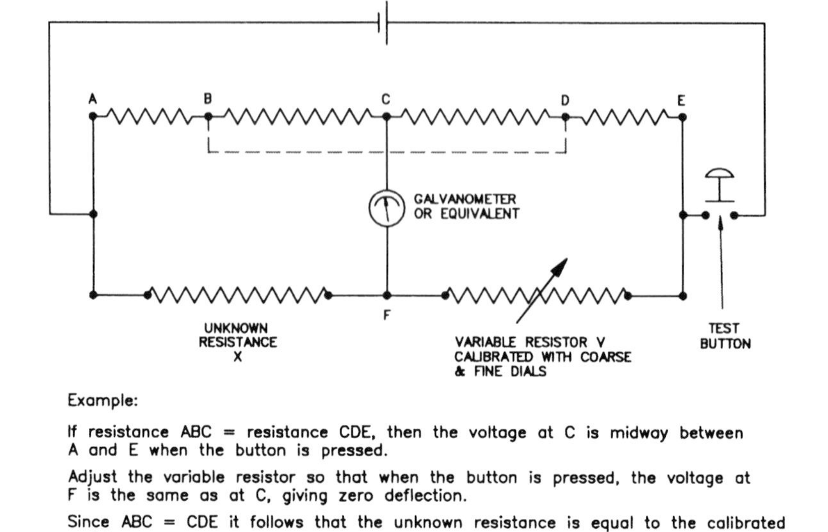

Example:

If resistance ABC = resistance CDE, then the voltage at C is midway between A and E when the button is pressed.

Adjust the variable resistor so that when the button is pressed, the voltage at F is the same as at C, giving zero deflection.

Since ABC = CDE it follows that the unknown resistance is equal to the calibrated resistor.

By connecting the galvanometer to B or D, balance is obtained when X + V = AB ÷ BCD (or ABD ÷ DE); conveniently 1 ÷ 10 or 10 ÷ 1 by choosing AB, BC, CD and DE to suit.

Fig. 7.5 Another form of the Wheatstone bridge to measure resistance

7.1.7 Potentiometers

A potentiometer (often called a 'pot' for short) is a resistor with an additional offtake point, either fixable or more often a wiper arm, which can be set at a required position, for the purpose of obtaining a voltage of some fraction of the total voltage across the whole resistor. It is not exactly a fixed fraction since the current divides at the offtake point so that there may be a small change due to heating, depending on the relative currents in the main and branch line after the offtake. It must be remembered that large currents in small devices may cause excessive heating.

A potentiometric voltmeter uses the principle of producing an adjustable voltage in a similar way and comparing it with the unknown voltage. It is connected to a stable voltage source, generated electronically or by a battery set up against a standard cell. The principle is shown in Fig. 7.6a where a bare slide wire is connected to a known voltage source. The unknown voltage is connected through a sensitive centre-zero meter to a moveable pick-up. The pick-up is placed on the slide wire at some point and the button is pressed momentarily. According to the deflection, trial and error is used to find the null point, when pressing the button produces no deflection. Then, assuming the wire to be uniform, the unknown voltage is in a ratio to the known voltage equal to the relative length of wire up to the probe divided by the whole length. The current in the wire is quite small and constantan would normally be used for this. The important feature is that the unknown circuit only takes or gives current during the balance-finding.

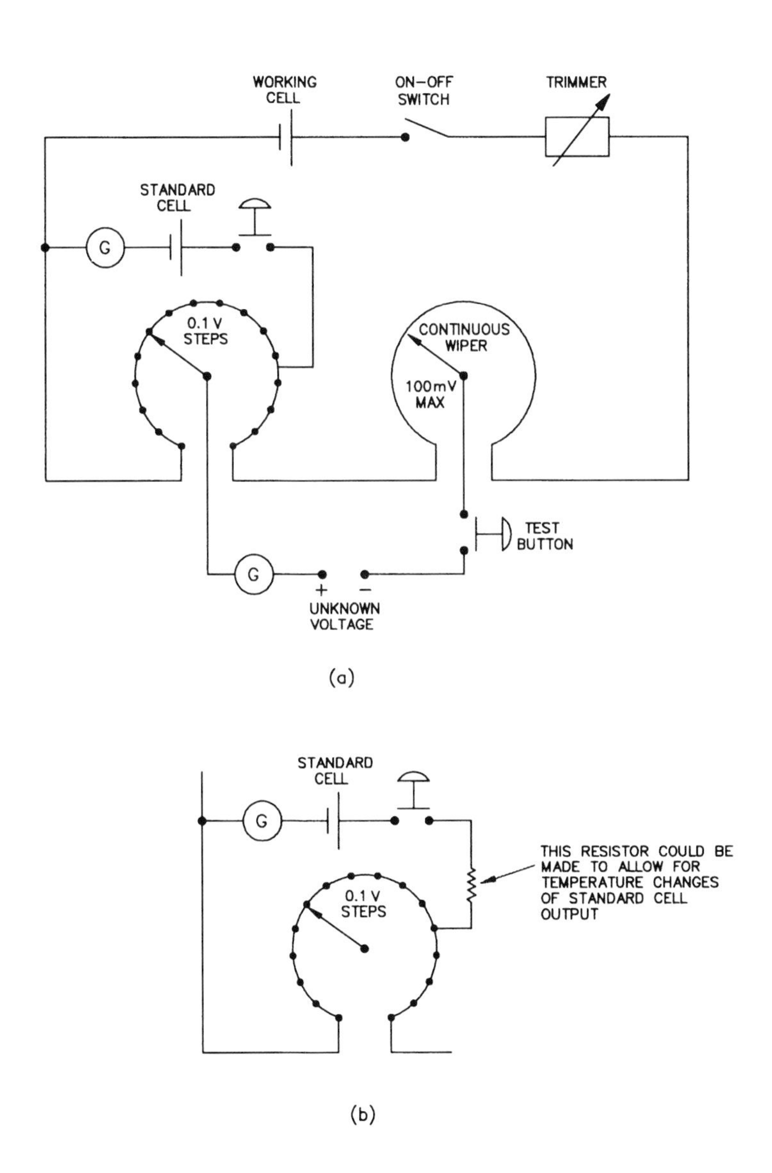

Fig. 7.6 The potentiometric voltmeter

A sensitive workshop potentiometer useful for thermocouples and similar work is shown in Fig. 7.6b. It uses a 1.5 V dry cell and a Weston standard cell; the Weston cell has a no-load voltage of 1.01876 V at 15 °C, 1.01837 V at 25 °C, etc. The dry cell is the working battery and since it does not have a completely steady voltage it is corrected from time to time using the standard cell as described below.

Instead of the simple slide wire there is a series of accurate resistors chosen so that when the selector switch is set, a stated voltage per step is taken off; this switch may produce steps of 0.1 V. The fine adjustment is by a wire-wound resistor of quite possibly 1000 turns, of total resistance equal to one step of the coarse selector. This is picked off by a wiper arm whose dial has appropriate graduations. The test button may contain a protective device against excess current passing through the galvanometer.

Since the working cell output may vary, a trimmer resistor is included in its circuit. Pressing the standardizing button connects the standard cell momentarily to the main wire at the point where the battery current gives the voltage matching the standard cell. It is possible to arrange for the working galvanometer to be switched in by this action but here a separate galvanometer is shown for simplicity. It is important not to press both buttons at once; this can be arranged by using a two-way break-before-make switch.

7.1.8 Electronic voltmeters

These meters resemble the potentiometer device of Section 7.1.7 but use electronics to perform the balancing as well as providing the reference voltage. They find the incoming voltage usually by frequent step-wise scanning. If the output is digital, the last digit has a disturbing habit of changing its value. This can be over-ridden by the user but no doubt future meters will be made more intelligent so that they reduce the number of changes per second if there is no rapid drift. There can be some harm in excessive scanning since scanning produces a current in the unknown load, which could affect the measuring set-up. Some meters cause currents during searching of as high as 2 mA. These are discussed elsewhere in relation to temperature measurement (Chapter 5) and effects upon strain gauges (Chapter 4).

7.1.9 Oscilloscopes

An oscilloscope is the most versatile, often the only, way of displaying changing phenomena. The catalogues show several pages of the traditional cathode ray oscilloscope, with varying facilities, at prices rising from ten times that of a good voltmeter. There are both analogue and digital scopes, the digital versions becoming substantially more expensive for high-frequency work. On an analogue scope an electron beam is made to sweep horizontally across a cathode ray tube by means of a sweep generator or time base circuit. The instrument can perform this at regular set intervals, or set it off with a trigger pulse. The beam is deflected vertically by a voltage signal. This can come direct from one of many transducers which respond to position change, pressure, speed, acceleration, force, temperature, proximity of some object, optical signals, etc. The oscilloscope includes amplifiers to increase the signal, or potentiometers to reduce it.

Many of the digital oscilloscopes can hold (store) a trace, using modern high-speed digital scanning. The trace can then be photographed or measured and sketched; moreover the digital information can be fed out and processed. The top of the range is the fully digital oscilloscope; the wave form being studied is scanned at such frequent intervals that the continuous observation provided by the cathode ray method becomes unnecessary. Modern digital scopes are provided with on-screen cursors to enable waveforms to be measured accurately, and often calculate frequency, rise time, etc.

If an oscilloscope seems to give mis-information this may sometimes be due to the trigger which starts the cycle. For instance in following the pressure cycle in a piston engine the trigger may be optical, when a white line on a dark disc sweeps past a photo-cell. If there is an oily streak on the disc this may masquerade as another white line and give an occasional mis-timed start. On a spark ignition system a spark plug lead trigger was used which insisted on giving two P–V curves displaced by about 20°, although the reason for this was not discovered.

If an oscilloscope is used as a speed counter, e.g. with a white line as noted above to give a blip once per revolution, then an error could show up as one or more blank sweeps without a blip. Normally one would be operating with a number of blips per sweep, in which case a missing or intermittent blip would be easy to spot, whereas on other pulse counters the error would not show up.

7.2 Transducer outputs and instrumentation

The transducers referred to in this book give an electrical output in response to a physical input which may be analogue, for example temperature, or digital such as a toothed wheel and opto-switch picking up, say, 60 pulses per revolution. The change in output is usually roughly linear with respect to the change in input, but may need some correction to linearize it. The output may also be very small, and need to be increased to transmit it over some distance to a convenient point where it can be displayed, recorded, or used for control purposes. It may then need to be converted, for example an analogue output to be shown on a digital display. Errors can be introduced into the signal either by the measurement system itself or from external sources such as electromagnetic interference. This section describes the transducer output, what happens to it downstream, and suggests ways to minimize or account for the errors.

7.2.1 Transducer output – static characteristics

A thermocouple will give a small electrical output when the measuring junction is subjected to a temperature different from that of its reference ('cold') junction (see Section 5.3). The output varies roughly linearly with the input, but correction tables are available, and thermocouple measuring instruments have corrections built in. A strain gauge varies its resistance with strain; in order to get an electrical output a stable external power source and circuitry is introduced (Section 4.2) .

The magnitude of the change in output relative to the change in input is known as the sensitivity. The range of a transducer refers to its maximum and minimum values; in the case of a thermocouple its input range might be 0–1000 °C, whilst its output range is 0–41.3 mV. This does not necessarily mean the transducer won't operate outside this range, but the specified error band may be exceeded, or physical limits such as the softening temperature of the sheath material may be reached.

A number of errors may apply to the output from the transducer before any further processing is done. Examples of these are non-linearity, hysteresis and limits of resolution, and more details of these are given in The Appendix. Environmental effects such as ambient conditions or the supply voltage can also affect the output. Often these are too small to be worth quantifying individually, and are collectively referred to as an 'error band' for the transducer.

The errors described above are systematic. Systematic errors will show up on the same side of the true value, and can often be calculated and the measurements corrected accordingly. Intelligent instruments can do this automatically. Analogue type compensating elements can be included to help linearize the output of non-linear transducers, such as a deflection bridge for a thermistor.

When the environmental conditions affect results it may be possible to isolate the instrument, or remove the effect. Matched strain gauges giving tensile and compressive strains, e.g. either side of a beam in bending, are connected in a bridge such that the two resistances subtract, giving twice the strain due to bending, and the effects of ambient temperature change cancel out (see Section 4.2.5). High gain negative feedback is used to reduce environmental effects or non-linearity. More details of this are given in Bentley (**22**).

Other errors can be random, e.g. caused by variations in the way different observers read a scale, and should be spread equally on either side of the correct value. They can then be averaged, or described in terms of the probability of being within, say, 95% of the correct value. This is also known as a level of uncertainty.

7.2.2 Transducer output – dynamic characteristics

The previous section refers to the output of a transducer in response to a static input value, known as the steady state response. In practice the input value is likely to change, and the transducer may take a finite time to respond. This is unimportant if the response time is much less than the time over which the measurement is required. However, the dynamic characteristics become important for high-speed work such as cylinder pressures in IC engine testing (Section 4.4), and impact testing (Section 8.6).

In the case of a sheathed thermocouple measuring a change in temperature, it will take some time for the heat to conduct to the couple itself and modify the output. The response time of the circuitry is, in this case, negligible by comparison. Improvements can be made by a smaller sheath size to give lower thermal mass, or if appropriate using a bare couple or earthing the couple to its sheath. Typical response times for sheathed thermocouples might be 5 seconds for a 6 mm diameter insulated thermocouple at 430 °C, as opposed to 0.04 seconds for a 1 mm diameter earthed couple at the same temperature.

The transient response of a thermocouple can usually be modelled using a zero-order (i.e. instantaneous such as a potentiometer) or a first-order differential equation, although it is sometimes necessary to go to a second or even higher order. It is then possible to predict the output at a given time in response to a change in the input signal. A time constant t may be quoted, which is the time taken for the output to reach 63% of its final value in response to a step change in the input.

Often the response of a transducer is modelled by using the 'transfer function' or ratio of output to input. The transfer function could be in terms of the ratio of magnitude or of frequency. If the transfer function is in the frequency domain, it is the ratio of the Laplace transform (which converts from time into frequency) of the output to the Laplace transform of the input signal, which makes the calculations easier to handle when more than one element is involved. It may then be possible to dynamically compensate the signal. More details of this are given in Bentley (**22**).

7.2.3 Signal conditioning and processing

Signal processing can be done by analogue or digital means; analogue being normally cheaper. Digital is often more accurate, although it can give false confidence in the readings or introduce problems such as aliasing, described in Section 11.1.5. Signal processing can consist of signal amplification, attenuation, linearization, removal of bias (i.e. an offset in the output signal) or filtering to remove unwanted frequencies, e.g. 50 Hz interference from a mains supply. Signals can be mathematically manipulated and converted from one form to another. Digital signal processing is normally conducted by an 'intelligent instrument' which corrects for known systematic errors in the measurement system, and analogue signals have to be converted to digital for this to be carried out.

Signal conditioning is processing done to convert the output of a sensor into a more convenient form for transmission or further processing, normally a current or voltage of chosen amplitude or frequency range. A current output has the advantage that signal attenuation due to the resistance of the conductor is minimized.

Thermocouples are often connected to a thermocouple transmitter, which converts a rather small, slightly non-linear input voltage to an equivalent 4 to 20 mA output current for onward transmission. The transmitter is externally powered and measures a voltage drop across a resistor in the thermocouple circuit. It measures the local temperature and applies cold junction compensation. A minimum current is needed to power the transmitter, so if the current drops to zero it is apparent that the circuit has been broken.

For measuring high-speed rotation where space is limited, one solution is to use a fibre optic sensor head consisting of two fibres, one of which transmits light from a photoelectric sensor/amplifier. This then reflects off a target at, say, once per revolution and the other fibre picks it up. The photoelectric sensor converts the reflected light into an electrical pulse and amplifies it for transmission as a digital signal. With strain gauges a bridge circuit is used to convert the change in resistance into a voltage output as described in Section 4.2.3. An a.c. bridge circuit can also be used for capacitive and inductive sensors, e.g. proximity sensors.

With piezo-electric transducers (Section 4.5) a charge amplifier is used to convert a small charge such as a few hundred picocoulombs to a voltage output, normally in the +/–10 V range. This contains adjustments for input scale or range, a low-pass filter, amplifier, and power supply. Because the charge is so small the input impedance of the instrument has to be very high, and the signal is highly susceptible to losses in the junctions and interference. Special cables are used to minimize these effects.

In order to avoid external interference and drift of amplifiers, the signal may be modulated for transmission. One way of doing this is to convert the signal to a.c. for transmission, then back to d.c. at the display point. Either the amplitude or the frequency of the signal can then represent the magnitude of the measured variable, frequency modulation being less susceptible to noise and more easily converted to a digital output.

7.2.4 Signal transmission, remote sensing, noise, and interference

Often the place at which a measurement is to be displayed, output, or used for control purposes is at some distance from the point of measurement, which can lead to various problems. The main one is electrical noise or interference being superimposed on the measurement signal.

The author was having trouble calibrating the load cell on a dynamometer used to measure the power output from a small gas turbine. A young apprentice was working nearby on an engraving machine; when he was called on to assist, the displayed signal rose by 8% and became steady. When he went back, so did the signal. This was because when he was assisting the calibration the engraving machine was switched off, and ceased to cause interference problems. The problem was solved by adding a low-pass filter at the input to the strain gauge amplifier.

Electromagnetic interference often affects instrument signals, particularly when operating with very sensitive instruments close to equipment that produces a lot of electrical noise, such as thyristor drives for a.c. motors which produce high-frequency spikes. Drives for d.c. motors also produce noise, but at lower frequencies than a.c. drives, as do solid state relays and other equipment with high inductance or capacitance, such as induction heating equipment. The E.E.C. electromagnetic compatibility (EMC) regulations are designed to eliminate radio frequency interference emissions from electrical machines, and to ensure that these machines are immune to such radiation from external sources.

To help limit interference problems, signal and power cables are routed as far as possible from one another, sensors such as load beams are electrically insulated from other equipment, and shielding is used where necessary and appropriately earthed. Low-pass filters are used to reduce high-frequency noise in signals, and the signals themselves are amplified as near to the sensor as possible to reduce the signal-to-noise ratio on the cabling. Sometimes the problem itself can be reduced by using, for example, d.c. instead of a.c. motors and drives. 'Mu' metal is useful for mechanical shielding of electrically 'noisy' components.

On a rolling contact friction and wear tester, interference from the a.c. motor drive affected the speed setpoint, causing the motor to run overspeed at high torques. Luckily this was observed immediately as the speed was being monitored. The problem was solved by fitting a low-pass filter across the input terminals to the drive.

Interference occurred on a tumble and swirl rig, where an a.c. fan was used to circulate air through a dummy cylinder head of an IC engine at a known pressure with the valves set to various positions. The tumble or swirl of the air in the dummy cylinder was then measured. Some very sensitive pressure gauges were used in the experiment, to monitor the pressure in the ducting. These were affected by noise from the a.c. drive to the fan motor. This was solved by more effective shielding of the signal cables and re-routing of earth wires.

Another problem that arose with this type of pressure transducer was in the signal conditioning unit. It had been built in-house for some less sensitive pressure transducers, and when used with these gave large errors. This was because the interface did not have a high enough input impedance; it was only 100 kilohms, and an impedance in the megohm region was required for these transducers.

Other sources of noise and interference include a.c. power circuits, solenoids switching, fluorescent lighting, radio frequency transmitters, and welding equipment. The interference occurs by inductive or capacitive coupling, or sometimes by having earths at slightly different potentials. Section 7.2.3 suggests ways to reduce problems due to interference by conditioning the signal before transmission. Shielding of signal cables and/or sources of noise is also used, and twisting pairs of wires together to cancel out induced voltages. Introducing physical distance between them is used, as mentioned above. Differential amplifiers subtract the interference voltage picked up by a second wire from the total voltage in the sensor wire. Filtering removes chosen frequencies. Sometimes the output is just smoothed, which is useful for random noise but for interference (deterministic) this can add an offset to the measured signal.

The use of optical fibres as a transmission path for signals avoids electromagnetic interference and reduces signal attenuation, although care must be taken at the amplifier input and output. It gives a fast response time, at somewhat increased cost. Section 4.6 gives more details.

When a measured signal has to be processed, perhaps transmitted over some distance and then utilized, there can be a significant effect on it. One element in the system can affect the previous one, for example by drawing current. Usually these effects are insignificant compared with the value of the signal and the accuracy required. Often they can be kept small, for example by using transmission wires of adequate size to avoid voltage losses, and by using a high load impedance when measuring voltage drop, or a low impedance when measuring current.

When the electrical loading effect is important it can be quantified using the Thevenin and Norton equivalent circuits. In the Thevenin theorem a network of linear impedances and voltage sources is replaced by an equivalent circuit consisting of a single voltage source and series impedance. For any load connected across the output terminals, the current and voltage drop can then be calculated. The Norton circuit consists of a current source and parallel impedance, and can be dealt with similarly. The method is useful to calculate the effects of multiple elements downstream of the transducer. Details and worked examples are given in Morris (**23**).

Chapter 8

Measuring Properties of Materials

This chapter outlines the various ways in which materials can be tested for strength, and some of the pitfalls which can be met during such testing. It also gives a rough outline of material behaviour, particularly that of metals as most of the tests are concerned with these. More details can be found in Higgins (**24**). Two case studies are included, one concerning impact testing of metals, and the other fatigue testing of metals with and without surface treatment.

There are a variety of ways in which a material or component can fail, and the different testing methods are designed as far as possible to simulate these modes of failure and discover more about them, generally to avoid failure of the component in service. Failure can imply actual fracture, excessive deformation, excessive wear, etc., indeed any failure to perform the intended function. In the context used here, failure usually means fracture, due to excessive stress being applied. The uniaxial tensile or compression test is generally considered the most basic and the data it gives can be applied to cases of steady tension, bending, and torsion, at room temperature. With some trouble a tensile test can be performed at other temperatures also.

Other important tests are for localized hardness, long-term loading at elevated temperatures, brittleness under impact, and, very importantly, fatigue tests. A material may fail at well under its static strength when subjected to continually varying stresses causing growth of cracks.

Another failure is by brittle fracture without impact to set it off, at low atmospheric temperatures. There are no definite tests for this property but the liability and immunity of various materials is known. The first well-known failure was a large tank full of molasses. Low temperatures tend to increase the risk greatly; metal mountaineering equipment such as carabiners should be made of material resistant to low temperature brittle fracture. Many grades of steel are subject to the effect, although failures are rare. During the Second World War the all-welded 'Liberty' ships replaced riveted versions for speed of production, and crack growth was no longer arrested by the plate boundaries or the rivet holes. At low temperatures the ships would break in half. One gentleman told the author he was appointed on these ships during the war, to chase cracks across the deck with a drill, in order to blunt them in time. It is sometimes said to be associated with fatigue (fluctuating loads) but tankers and bulk carriers have also been known to break during relatively calm weather.

8.1 Metal behaviour

Before going into details of the tests a brief discussion of fundamental behaviour seems appropriate. If we imagine failure as overcoming the atomic forces directly we might expect that a piece would extend perhaps a fifth of the atomic spacing; in other words the tensile strength would be a fifth of Young's modulus and the break would be sudden. There would be no bending, the material would be difficult to work, perhaps like cutting glass or diamonds. The nearest approach to these properties are found in boron fibres, carbon fibres, and certain polymers.

The usual properties obtained in practice are due to the multitude of crystals. Before metals were examined microscopically it was thought that crystallization was abnormal and weakening. This error was due to coarse fracture surfaces in brittle metals. When specimens, gently polished but not buffed at high speed, were examined the general crystallinity was revealed, helped by etching the surface.

8.1.1 Crystal planes and dislocations

When metals solidify from the melt, crystals grow mainly from the coolest edges until they meet from opposite sides, though some may grow from within. With fast cooling, the crystals are many and small. Any subsequent working breaks them up further, though the atomic forces are so strong that no cavities result unless they have formed during the solidification. It must be said that in recent experiments metals have been chilled so rapidly that they set in amorphous, glassy form. The properties of these are interesting but too new to be discussed here. One reason why metals can be shaped without breaking is that in the crystals the pattern of atoms is not entirely regular; there are occasional dislocations similar to creases in a carpet laid in a room with irregular edges. The name dislocation does not imply structural weakness but rather a form of softness. The dislocations can be pushed along by overcoming only a limited number of atomic forces at a time. They do not seem to cross easily from crystal to crystal; accordingly, a fine crystal structure tends to go with higher strength and lower deformability.

The sliding of layers, accompanied by some rotation of the crystals, is called shear flow, from its resemblance to scissor action though without actually cutting. It has long been accepted that shear flow is the fundamental mode of metal deformation. Even in tensile testing (pulling) this shear is the main action and can sometimes be observed directly, at 45° to the pull. Under suitable magnifications dislocations can be observed. Work-hardening causes them to collect near the grain boundaries.

In recent years it has become feasible and worthwhile to produce single crystal components substantially free from dislocations, or directionally solidified, e.g. for some aerospace applications. This results in exceptional properties, particularly creep resistance (see below). Seemingly, the dislocations have further to go before they affect the component.

During large shearing motions ductility is ‘used up’, the material work-hardens (strain-hardens). Metal alloys with a second constituent of appreciably different atomic size, e.g. carbon in iron, are usually stronger and less ductile than the pure metal concerned, due to a roughening or pegging effect on the crystal planes; this makes the dislocations less ready to flow.

The atom pattern (crystal lattice) has several forms, depending on the way the atomic forces are arranged. The main forms are described as face-centred cubic, body-centred cubic and close-packed hexagonal. An alloy may be uniform (one phase) or may have two different phases intermingled. This does not affect behaviour greatly except with iron which has the two different forms of cubic lattice, one at high temperature and one at low.

8.2 Effects of heat

Heating (annealing) normally reduces the effect of work-hardening, by growing larger crystals or at least by easing the mismatches at the boundaries. The familiar hardening of steel by rapid cooling from above a certain temperature is due to the unusual property of iron noted above, in that it has two crystal lattices, relating to different temperatures. Transforming from the high- to the low-temperature phase happens in the solid state when cooling through a certain temperature band. This transformation creates fresh crystals. If a suitable amount of carbon is present it gives a hard brittle state which needs moderating (tempering) by limited heating. The transformation temperature can be changed, or the transformation suppressed altogether, by certain alloying elements. For instance, austenitic stainless steel stays in the high-temperature phase down to room temperature but can easily suffer loss of stainlessness after local heating, presumably having been partly transformed. It can be stabilized against this; if intended to be welded, etc. the composition includes over 2% titanium and/or other elements.

Another way of making metal harder by heat is precipitation. The alloy is made soft and homogeneous, usually by holding at a certain temperature for some time; this puts the pegging constituent into solid solution. This is followed by chilling quickly to hold it there, which keeps the metal soft. Subsequently a second, lower heating precipitates the pegging phase. In some systems it can precipitate out cold, during working, or by itself, one example being aircraft rivets going harder and stronger after setting, or inadvertently during storage. Deliberate precipitation is the common treatment with high-strength aluminium alloys. It can apply to other metals too; for instance low-alloy precipitation-hardening steels are used increasingly, giving good strength-to-cost ratios (and saving weight in transport applications).

8.3 Static tensile and compression tests (actually slow-moving rather than strictly static)

Whilst in ordinary speech the terms stress and strain are used interchangeably, in engineering the word stress is reserved for force per unit area. We distinguish tensile, compressive, and shear stress. Shear stress can be linear or in torsion (twist). The term strain, tensile or compressive, is used for change of length per unit distance. Shear strain means lateral displacement per unit distance measured at right angles, i.e. the tangent of the angle of distortion.

Static tensile tests are commonly used for metals at room temperature. For service at higher temperatures, as in engines, strength data are available from books and manufacturers, including data on long-term properties. Polymers generally have much lower strength than the main structural metals. Moreover, many polymers creep at room temperature, so time and temperature come into the equation. Composites such as bonded carbon or glass fibre are much stronger, and have improved creep resistance. Ceramics generally have enormous yield strengths, but a very low fracture toughness, so that for a tensile test at room temperature they tend to fracture before yielding. A compression test, or preferably a hardness test, is more appropriate.

Routine tests during production may be carried out on plain round or flat bars, to detect inconsistencies. However, for most purposes it is necessary to use recognized test pieces and procedures. BS EN 10002 (**25**) shows considerable freedom of specimen shape and size: the most important provision is sufficient length to allow metal shear flow to develop adequately. Figure 8.1 shows schematically a small tensile testing machine and a specimen. Ends must be smoothly enlarged to ensure that the effect of gripping does not falsify the results; in flat material (as shown here) the enlargement need only be in one plane. The uniform length (gauge length) must be at least 5.65 $a^{0.5}$ where a is the cross-sectional area. This is five times the diameter in a cylindrical specimen. When testing wires enlarged ends are not feasible; wire specimens have to be longer and may be merely clamped at the ends. An extensometer is shown schematically in the figure; this ensures that any distortions in the end sections do not influence the calculated results.

In the testing machine the specimen is pulled slowly while a weight-and-lever or strain-gauge-controlled mechanism is used to indicate the force exerted. In many cases the machine uses instrumentation which automatically produces a load–extension diagram. The tensile test machine normally operates by applying a constant strain rate to the specimen, for example using servo-hydraulics with feedback control of position, or for lower loads by moving the load carriage at constant speed with a ball screw and servo-motor arrangement.

The first part of the test, in metals, produces a straight line, confirming that the strain is proportional to stress, i.e. Hooke's Law. (Nowadays one may consider this law rather obvious, but in Hooke's era common experience was with ropes and leather, showing non-linear behaviour.) During this stage the process is elastic (virtually reversible); unloading produces almost identical lines and the specimen returns to its original size. The cross-sectional area diminishes slightly under stress (the Poisson contraction) but it seems to be usual tacitly to ignore this in the stress values.

Beyond the elastic range, many steels show a distinct yield point, as in Fig. 8.1, extending at least a little without demanding an increase of force; in many cases the force exerted actually diminishes at this stage. This is the onset of permanent deformation (plastic flow). With more extension, the force usually starts rising again; the recovery point is called the lower yield point, as distinct from the first (upper) yield point. The shear planes mentioned above can sometimes be seen at this stage. Harder materials may not show the upper and lower yield points clearly.

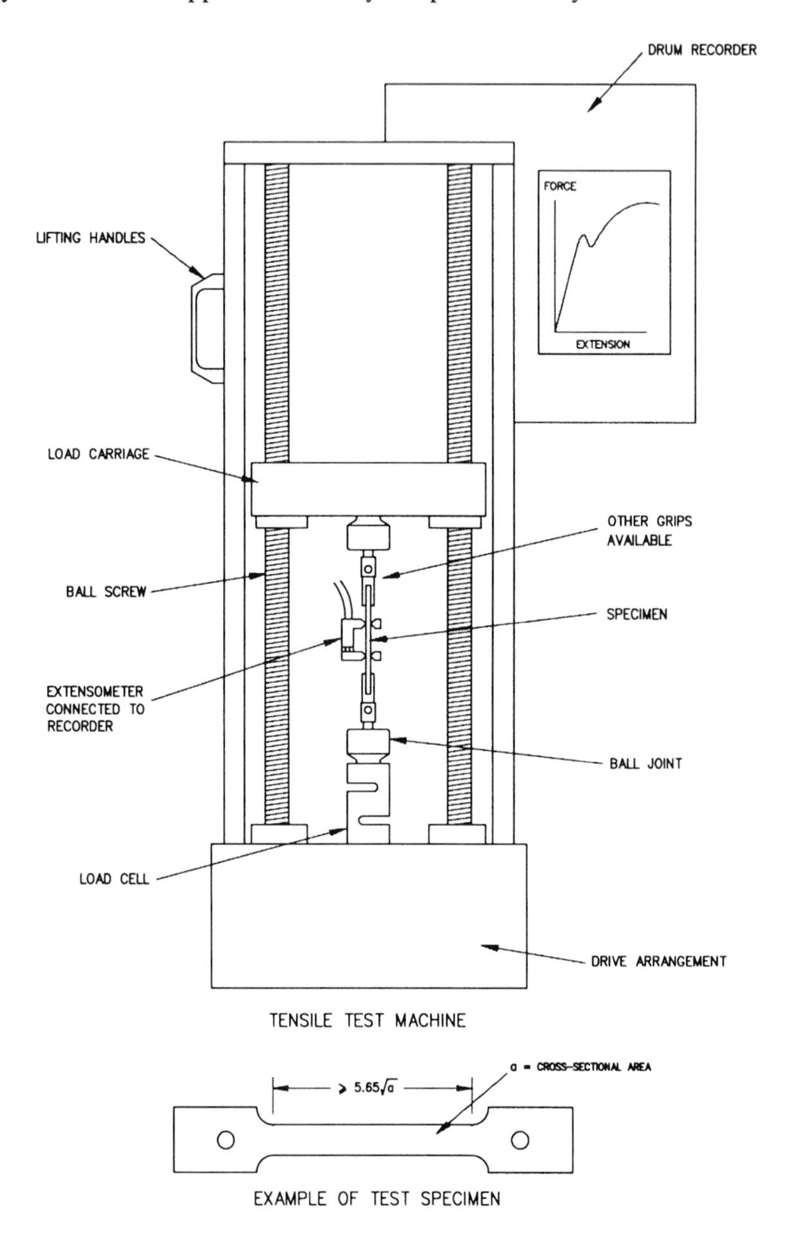

Fig. 8.1 Schematic of a small tensile test machine

Eventually the force reaches a maximum; soft (ductile) materials do not generally break at this stage but get longer and thinner, exerting a diminishing force. Eventually a neck forms and the specimen breaks. By fitting the two parts together, the elongation can be measured. Harder materials, including harder steels, simply show a slow departure from linear behaviour, accompanied by some permanent stretch, but without the turn-over. When the yielding is not definitely clear, such as with relatively hard material, it is characterized by drawing a line on the diagram parallel to the elastic line, offset by say 0.2 % strain. This line meets the curve at what is called the 0.2 % proof stress.

It is worth noting that the strength value normally quoted called the ultimate tensile strength (UTS) is the peak applied force divided by the original cross-sectional area, not the actual cross-sectional area, i.e. it doesn't take into account the area reduction due to 'necking'. The UTS is a practical value rather than an academic exercise. It should be explained that engineering designers use the UTS to make the shape and size of their components fall well within the capability of the intended material; they use a factor of safety. In this they are assisted by experience (not only their own) and by codes of practice, often compulsory in the appropriate field such as bridges, aircraft, pressure vessels, etc. Some codes of practice speak of design stress as compared with the UTS rather than a numerical safety factor. The design rules also take note of the yield point or proof stress, and of the endurance limit in fatigue (see below) if appropriate. The percentage elongation is mainly used as an assurance of ductility.

The graph is potentially confusing, as the breaking strength appears to be less than the maximum tensile strength when in fact it isn't. The relevant properties are the reduction of area and the true stress at fracture; these are mainly of interest for further working of the material, not directly in design work.

Compression tests are sometimes required, e.g. for concrete and grey cast iron which tend to be stronger in compression than in tension; most tensile testing machines have a facility for such tests. Compression tests are usually confined to brittle materials; ductile materials show an initial yielding at much the same stress as in tension but thereafter the test is seriously affected by the amount of friction at the end faces which restrains flow and results in barrelling. The specimens are necessarily kept short to avoid lateral buckling.

Compression tests are used as a convenient quality control check on some materials such as plastics, since the specimen can be an easily produced cube or short cylinder. The presence of solid lubricants such as PTFE in the plastic can lead to apparently low values of compressive strength, due to the reduction of friction effects at the loading platens. Repeating tests with the platens alternately greased or clean can demonstrate the substantial effect of friction, particularly with low-modulus materials such as plastics.

8.4 Creep (and relaxation) tests

Creep is defined as time-dependent plastic deformation, and in most materials starts to occur at absolute temperatures above 0.3 to 0.5 of the material's melting point. This has two important consequences; firstly that at constant stress, strain accumulates with time, for example turbine blades can extend and contact the turbine casing, and secondly at constant displacement, stress relaxes with time, so that at high temperatures, bolts may need re-tightening regularly [see reference (**26**) for further details]. Hence the need for a relaxation test. Some materials, for example glass or pure lead (without a strengthening alloy element), deform under their own weight over time, even at ambient temperatures, e.g. a glass thermometer becomes inaccurate over time as the bulb changes size through creep.

In a creep test, often a constant load is applied to the test specimen. This may be through weights or applied through means such as the servo-hydraulic or ball screw arrangements described in Section 8.3, only with feedback control of load. Extension is then measured against time, for a set temperature which is normally controlled using an electric furnace situated around the test specimen. The higher the temperature, the faster the material will creep for a given load. The relaxation test is similar to the creep test, except that the reduction in force for a given extension is measured.

It is important to be aware that most types of creep testing machine use constant load as opposed to constant stress. In other words, as soon as the specimen starts to neck, the load is not backed off to compensate, meaning that failure will occur quicker than under constant stress conditions. The above examples make it clear that both versions are important.

8.5 Hardness tests

Hardness is the ability of a material to resist surface abrasion or indentation, so hardness tests are used for materials which are likely to be exposed to this. Traditionally, geologists have used the Moh's table to compare the surface hardness of a substance with a set of standard materials ranging from talc (1) to diamond (10). The surface hardness of the substance is determined by finding which standard substance will just scratch it. However, for more accuracy, and to be able to distinguish between different metal alloys, for example, the standard hardness tests were developed.

In this form of test local indentations are made in the material. The test machines are relatively small compared to tensile or creep test machines. They use a ball or a shallow pyramid indenter (Brinell, Shore, Vickers, etc.) and do not require elaborate specimens. As the test is relatively non-destructive, it may be possible to test the actual component rather than a separate specimen, providing that the test can be carried out on a non-critical region of the component. A constant force is applied for a short time. The deflection is not usually measured (except in tests on rubber, etc.), instead the width or diameter of the visible indentation is measured to give a value of force per unit area. This is usually the projected area, but in the case of Vickers hardness the actual surface area of the indent is used.

The action of such indenters does not crush the material but displaces it sideways and slightly upwards at the outer edge, hence the result is related to the material yield strength. The hardness if expressed in the same terms as tensile stresses comes out at 2.5 to 3 times the yield point, but of course it refers only to material near the surface, which may have been case-hardened or otherwise treated to have quite different properties from the general bulk.

Portable hardness testers working on the same principle are useful in identifying bar materials where the usual paint marking has been cut off or obliterated, and can also be used for checking large components or parts of large machines *in situ*, without extensive dismantling or cutting up.

Accurate results require a reasonably smooth surface on the test piece, in order to determine clearly the borders of the indentation. Rough machined surfaces, and fracture surfaces on broken components, are therefore more difficult to measure accurately. For this reason, and also to allow for random variability in the material, it is normal to take several readings and use the average value.

When measuring hardness of very thin hard surface layers, such as occur with nitrided or electroplated surfaces, the depth of indentation for the conventional hardness tester will mean that the result largely reflects the underlying material hardness, not the characteristics of the surface layer. For this situation, micro-hardness testers are used which work on the same principles but employ smaller indenters and much lower forces (down to 0.1 N), so as to measure the characteristics of a much smaller volume of material.

Note that the hardness tests are normally carried out at room temperature, whilst the material being tested may be used in an abrasive situation, will probably be operating at a higher temperature, and other conditions such as moisture content may be different, hence this test may be just the first port of call. Friction and wear testing is discussed in more detail in Chapter 9.

8.6 Impact tests

Impact tests are carried out to test a material's resistance to a sudden application of force to give an indication of brittleness; this kind of test often predicts the likelihood of brittle fracture better than a combination of high UTS and low elongation. However, the values cannot be used as design figures (as yet), only as confidence indicators.

In the early days of iron casting it was easy to attach a simple piece of wood to the foundry pattern so that every cast produced a rough test piece. This was important before it became known that quite small proportions of phosphorus, sulphur, or other elements could produce serious brittleness. The difference between disastrous brittleness and tolerable toughness became apparent with just a vice, a hammer, and a little experience. From this developed, eventually, the standard Charpy test described in BS EN 10045-1 (**27**), using a machine shown in Fig. 8.2.

There are two traditional pendulum type methods; the Charpy and the Izod tests. The notches are different but the main difference between these is that the Charpy specimen is mounted so as to fail as a simple beam in three-point bending, whereas the Izod one fails in a cantilever beam bending mode. The Izod type tests are normally carried out at room temperature as the specimen holding arrangement is integral with the machine. This is a disadvantage, as brittleness varies with temperature, and the Charpy tester is preferred for most applications.

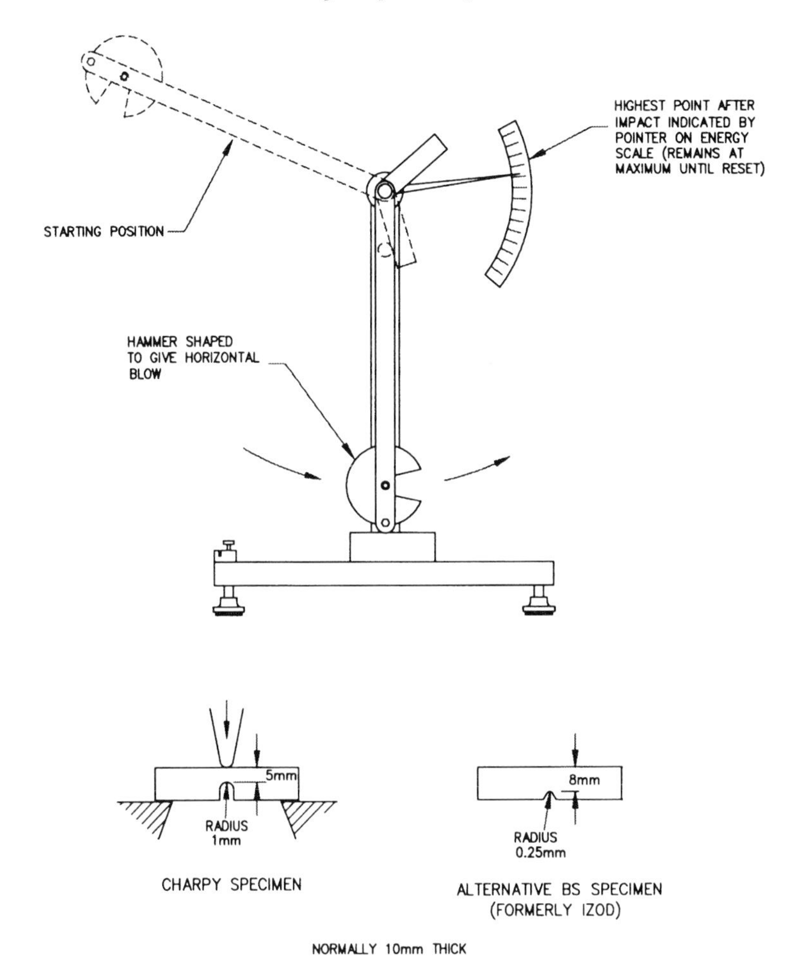

Fig. 8.2 Charpy impact tester and specimens

The design of the impacting element is fairly critical, to avoid it jamming against the broken specimen; something close to a knife edge is preferred. The test piece is laid between the jaws of the machine, then the swinging hammer is released from a height which will give an impact speed of 5 to 5.5 m/s (4.5 to 7 m/s is acceptable for pre-1983 machines). The available energy should preferably be 300 J, though smaller energies are accepted provided that the energy is quoted with the result. The hammer strikes the specimen and the energy absorbed is shown by the maximum height reached after the impact, indicated by a simple pointer which registers the top of the hammer motion. Other machines use an encoder fitted instead of the manually re-set dial, to give a digital readout of the maximum angle reached by the hammer. These are programmed to give the point of reversal of the pendulum as the maximum angle. Two types of test pieces are allowed, as shown. When quoting the energy absorbed in breaking, the distinction must be made which type was used; the form of description is laid down in the standard. This freedom may seem excessive for a testing standard but there is not much harm in it since the value is not used as a design figure, merely to discriminate between tough and brittle materials or the transition temperature at which a given material changes from tough to brittle behaviour.

The test is so quick that if a pre-chilled specimen is used there is no need to re-measure its temperature as it is laid into the machine. A number of steel types, tough at normal temperatures, can become brittle at low temperatures. This subject is by no means fully explored since some breaks have occurred at temperatures where other very similar cases have survived. From analysing certain older data (**28**) in terms of energy absorbed per unit fracture area (J/mm^2) it appears that actual size, width for a given thickness, and hammer speed can affect the energy of fracture and to some extent the transition temperature at which toughness begins to diminish. The effect of width may be surprising at first but this is also found to be relevant in fracture toughness testing as set out in the relevant British Standard. The background to this effect may lie in the Poisson contraction, see Section 8.3.

Case Study

Charpy impact tester

Some Charpy testers are instrumented, for example with strain gauges fixed behind the point of impact on the hammer, in order to obtain more information about the failure than just the overall energy absorbed. For example, Fig. 8.3 shows two graphs of force against distance measured during the impact, the first for a ductile material and the second for a brittle one, and from these the energy absorbed just up to the point of failure can be obtained (shown by the hatched area on the graph).

One such machine was used for testing samples of materials used in the structure of reactor vessels for nuclear power plants. The Charpy specimens were inserted in the vessel during use, and were thus subject to temperatures and radiation representative of the vessel in service. These were then taken out during shut-down and the tests conducted. The results of the tests were then used to help determine the future life of the vessel and the temperatures at which it should be run.

For testing at different temperatures the specimens were cooled in liquid nitrogen down to –180 °C, or heated in an oven up to 400 °C, then removed automatically, and the test conducted within two seconds to minimize heating or cooling of the specimen.

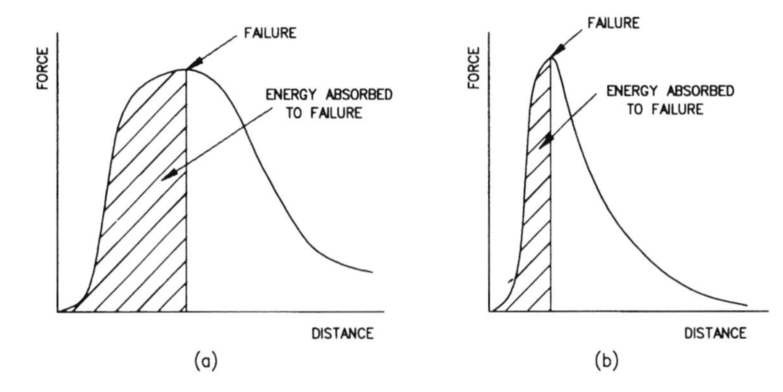

Fig. 8.3 Graphs of force against distance during impact for (a) a ductile material, and (b) a brittle material

The Charpy tester itself was instrumented as shown in Fig. 8.4a. The hammer consisted of a 'tup', into which a 'striker' was inserted and then screwed down (Fig. 8.4b). The striker had a conical hole in the rear, immediately behind the point of impact, into which were mounted four strain gauges connected in a full bridge. The wires were then led off through a tube welded on to the back of the striker, which passed through a hole in the tup.

One problem which arose with this method, was that any bending moment exerted on the striker gave rise to a strain, which could be larger than that measured during the test. The weld at the back of the striker to attach the tube had been ground down, but some residue remained. This meant that the striker was not flush with the surface of the tup but sitting on the weld residue (Fig. 8.4c), so that tightening the screws exerted a bending moment on the striker that swamped the impact signal. To avoid this the tup was recessed very slightly to clear the weld residue.

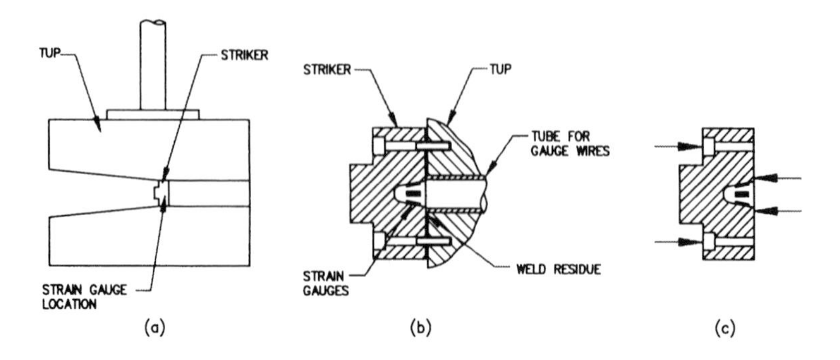

Fig. 8.4 Details of instrumented hammer showing (a) the strain gauge location, (b) the mounting of the striker, and (c) forces on the striker

A second problem which occurred gave rise to the graph form shown in Fig. 8.5a. The flat top to the curve was found to be caused by the striker being a tight fit in the tup, so that under a high normal load when the specimen tried to deflect laterally, it was restricted. This prevented the strain in the direction of impact so that the graph topped out. This was cured by removing about 0.05 mm from the tup to create a loose push fit. This also solved a third problem, which was the self-weight of the tup in the 'up' position inducing a bending moment in the striker and causing a zero offset of around 1.5%. This machine, due to the short time allowed from oven to test, was zeroed in the 'up' position.

However the greatest problem experienced was that of resonance on impact, giving a graph as shown in Fig. 8.5b. The period ΔT shown on the graph was found to correspond to the natural frequency of the tup when acting as a tuning fork. This was eventually solved by isolating the striker from the tup with a PTFE spacer, and PTFE washers under the screw heads. This meant that the tup had to be modified to ensure that the centre of mass remained at the point of contact, so the tup had to be machined back and the system re-balanced. The solution was then accepted as a standard procedure to avoid the tuning fork problem. The disadvantage of the PTFE was that the energy absorbed during the impact was slightly higher, although the results were still within the manufacturer's stated tolerance.

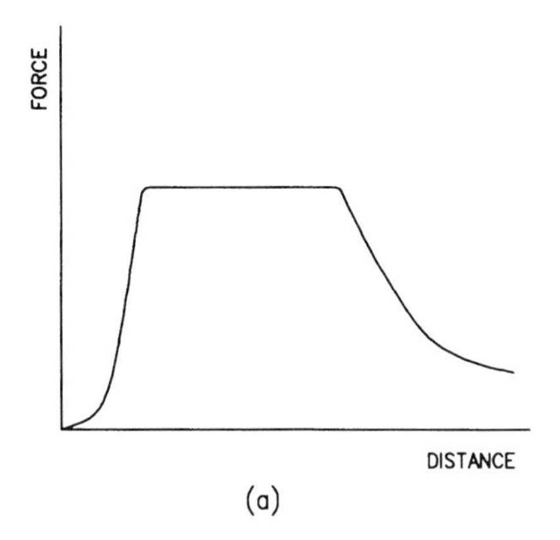

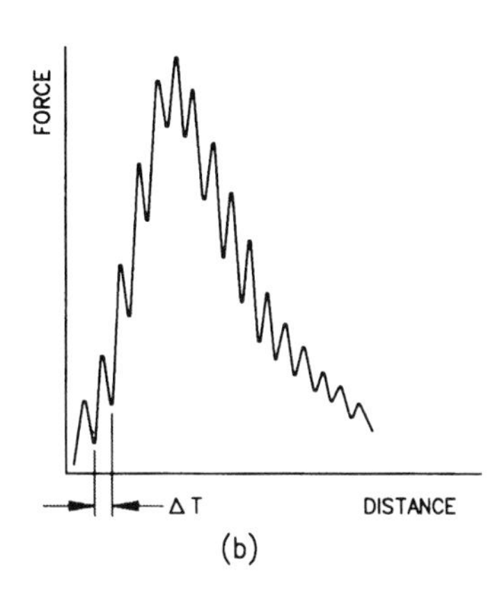

Fig. 8.5 Graphs of force against distance during impact showing (a) topping out of the graph due to lateral restriction; (b) resonance caused by the 'tuning fork' effect

The strain gauge arrangement was originally calibrated statically, but then the dynamic readings were found to be slightly slow, partly due to loss of the high-frequency components by a data acquisition system of a finite bandwidth (up to 100 kHz in this case). Calibration was done against the dial showing the overall energy loss, using known ductile specimens where the high-frequency losses were relatively small. For more brittle specimens a higher proportion was lost, and for materials such as ceramics it was found necessary to move to a much higher data acquisition rate, in this case 500 kHz.

Another version of impact testing is conducted by dropping a weight vertically on to a piece of material, via strikers of different designs. The machine consists of a vertically guided, free falling weight, and a rigidly supported anvil which allows the plate specimen to be loaded as a simple beam. The specimens may be notched, and tests conducted over a range of temperatures. In its most basic form, this test merely distinguishes between break and no-break conditions. However, the machine can be instrumented; the strain, or alternatively the acceleration, is then measured by a piezo-crystal mounted between the weight and the striker, and does not suffer from the tuning fork problem experienced in the Charpy tester.

8.7 High strain rate tensile testing

High strain rates occur in explosive situations; academically speaking it is desired to relate this to tensile test results.

It seems that if tensile stress is applied at high speeds the ultimate tensile strength is higher than in slow tests. The testing and analysis are both difficult, due to stress wave effects. For instance in one paper tubular specimens were subjected to a gaseous explosion. A single, steep up-and-down wave was detected, travelling very quickly along the pipe so as to amount to a suddenly applied force. The analysis was carried out on this basis, including overshoot past the static-level state and into the yielding region.

It was claimed that the strength of the material was almost double the static strength. However, it seems difficult to decide whether to calculate with the momentary peak pressure or with the after-pressure. It also seems fair to ask whether the detected wave was the gas explosion front or a stress wave in the pipe wall. The shear wave in the wall can be expected to advance much faster than the gas explosion front and be perhaps on the rebound before the gas pressure reached it. The point we wish to make is that the subject is difficult and perhaps too controversial for this book.

8.8 Fatigue tests

Fatigue seems to imply something related strongly to time; but time is at most a minor feature compared with the number of reversals, repetitions or fluctuations of stress. Sometimes the impression is given that a fatigue failure is like a sudden disease; indeed it is usually sudden, after cracking has gone on for some time, unobserved, finally leaving so little cross-section that the next load causes the break. There is always a reason for such cracking: either there has been a failure to design the component for the varying load, or the component has been misused in some way, or has suffered unforeseen effects from its environment, or undetected defects were present in the component. Often cracks propagate from stress raisers such as keyways, sharp radii or even machining marks. In some applications it is possible to monitor a structure's overall behaviour in terms of deflections or resonant frequencies, obtaining early warning of progressive cracking. Examples are known where an operator has detected a slight change in the sound tone emitted by a high-speed rotating machine, and has stopped the machine and discovered a rapidly propagating crack in the main shaft. A fatigue crack, in one case, was traced to the inspection mark made by the inspector's stamp.

To design safely for non-steady stresses it is vital to know the permissible stresses for repeated or unsteady loading, which are much lower than those for steady loads. There are two types of test: the easiest is to rotate a specimen and, through suitable bearings, impose a bending moment. At its simplest, a short cantilever test piece is placed in a rotating chuck (holder) and a weight is hung from a bearing secured at the free end. This tests only the portion where the stress is greatest. The diameter can be made tapered so that an appreciable length has uniform surface stress. Alternatively, a four-point bending set-up is used, with a region of constant diameter and constant bending moment.

Bending tests on small specimens are open to two criticisms; firstly, only the outermost layer is under full stress, secondly the stress diminishes from the edge inwards more rapidly the smaller the specimen. As Griffith (**29**) discovered, we should take note of material extending some distance from a possible crack, say by several grain diameters. The steep stress gradient in a small bending specimen makes it less relevant to a larger component; this size effect has been adequately demonstrated especially at points of local stress concentration.

The preferred method is push–pull testing; this gives uniform stress over the whole cross-section. Modern short, rigid testing machines broadly similar in configuration to that shown in Fig. 8.1, with automatic controls, can make steady or fatigue tests, with a wide range of programmes.

In early days very high-speed machines were favoured, to obtain quick results, since tests may cover 10 million cycles or more. Some difficulties were met such as inertia errors, temperature rise when cracking has started, and some doubt about whether the time factor was relevant. Wohler's original work (**30**) gave the specimens appreciable time, making stepwise turns during the loading cycle. It is necessary to test at a high enough speed to avoid strain rate effects, and slowly enough to avoid impact conditions.

The push–pull method is versatile since it can give cycles of full reversal or other variations. The range undergone has an enormous influence on the stress levels for a given survival. To describe the range neatly, the load ratio (also called stress ratio) is expressed as the lowest tension divided by the highest tension, compression counting as negative tension. Thus a figure of –1 means full reversal, 0 means repeated tension with zero compression. Other testing values often met are +0.5, or –0.5. For actual working conditions the designer interpolates within the data as needed.

From a series of tests on a material, at different stress levels, an *s–N* curve is produced, showing how the stress influences the number of cycles before fracture, having defined the particular stress ratio used, most commonly –1. In steel and some other metals it is found that at a stress which permits 10^7 cycles, there seems to be no limit to further life. This is called the endurance limit; usually to be specific it is called the plus-or-minus, or fully reversed stress endurance limit. In corrosive conditions there is a mutual aggravation effect called corrosion fatigue. Some recent work shows that for modern very clean steels with virtually no inclusions as used in rolling bearings and high-performance gears, the endurance limit occurs at higher stresses and a lower number of cycles, perhaps as low as 10^5 cycles.

For most non-metals, and many non-ferrous metals and alloys, there is no endurance limit but published data seem to run out after 100 million cycles. When testing polymers there is an additional complication, because due to their low thermal conductivities the energy absorbed by the test sample cannot be dissipated quickly enough, causing a rise in temperature of the test piece. This then decreases the fatigue life of the specimen.

In practice one mostly gets irregular loading and there is some confusing evidence on the effect of high-before-low or low-before-high loadings. At stress-raisers an early high load can produce favourable residual stresses by yielding some of the material. For instance if a concave corner is overstretched (yielded) in tension, then as the load is removed the adjacent structure squeezes the stretched region back into place, leaving it with a residual compressive stress. Under any later loading, of lesser magnitude, this material has less tensile stress by the amount of the residual compression. As may be expected this delays or altogether prevents the onset of fatigue cracks. Hence under such conditions, high-before-low can be very favourable. This is the basis of autofrettage in gun making. In a pressure vessel context it was investigated by Taylor (**31**).

On the other hand, with smooth specimens, it has been noticed (**32**) that unbroken specimens which had survived just below the endurance limit for the usual number of cycles could then be put back into the test at higher stresses and be found stronger, i.e. to have acquired a higher endurance limit. This was presumably a form of work-hardening.

Perhaps the chief precaution to consider in fatigue testing relates to residual stress noted above. When setting up a test, it may be tempting to apply a load just to check that everything is tightened up. An excessive check-load of this kind could so modify the conditions that the following results are invalid. This happened when developing the De Havilland Comet airliner. A test fuselage was auto-fretted before fatigue testing, just to make sure of integrity and sound connections, by people not conscious of the 'good' consequences. The specimen fuselage was then fatigue tested and found satisfactory, resulting in false confidence. Working aircraft were not autofretted to the same extent and failed in fatigue, leading to many deaths during airline service (**33**). Therefore, if a fatigue test result is obtained following any substantial prior loading it is essential that the fact should be included in the reported data.

It should be noted that prior loading is only one of several ways of inducing a local prior stress. Nitriding, case hardening with carbon or cyanide, shot peening, and surface rolling all generally produce a compressive surface stress. In most cases this is beneficial to fatigue performance but not invariably so; if the local working stress is compressive, the additional imposed stress may not be beneficial. For this reason, treatments such as shot peening are often applied selectively, avoiding areas where high compressive stresses are anticipated from service loadings. As for specimen testing, it is almost impossible to simulate a locally treated case correctly; the recipe of the treatment and the grain size all should be scaled to the right ratio; hence the safe course is to sacrifice a number of finished components for full-size fatigue testing. Fortunately such local treatment is not normally applied to major structures as a whole.

Case Study

Fatigue testing of shot-peened specimens

A series of tests were conducted to compare the mechanics of crack development in specimens with shot-peened and as-machined surfaces, as described in reference (**34**). The compressive surface stresses introduced by the shot-peening process are supposed to inhibit crack growth under fatigue loading. These residual stresses tend to diminish during the life of the specimen, and the study then went on to determine the effects of re-peening the specimens after partial fatigue. Most of the work was done by fatigue in four-point bending, giving a fully reversed stress cycle. The advantage of four-point over three-point bending is avoiding variations in bending moment along the specimen in the region where the crack growth initiates, except those caused deliberately by the design of the specimen. Otherwise different results would be obtained according to which way the crack grows.

The work was also backed up with some experiments using a uniaxial servo-hydraulic rig, giving a constant alternating stress over the cross-section. However, this loading method does not tend to show the advantage of shot peening over a machined surface finish, as the cracks are more likely to grow from inside, in fact sub-surface tensile stresses caused by the shot peening can actually make matters worse. It was necessary to machine a much tighter notch to initiate surface crack growth, and the final phase of crack growth was slower; the rotating bend specimen reduced in effective diameter with crack growth, rapidly increasing the stresses. The uniaxial specimens were much more complex to manufacture; the rotating bend specimens needed axial symmetry and were simply turned. The uniaxial specimens also had to have their ends threaded and coned to take reverse loading. They were milled square to aid crack detection and also had notches machined in the side.

The machined and shot-peened surfaces were evaluated prior to testing using Talysurf profilometry. The crack sizes were monitored firstly using a dye penetrant method, and also using the a.c. potential drop across the crack, giving an indication of the crack depth, although the results obtained with this method also vary with the crack shape. The most reliable results were obtained after the failure by looking at the fracture surface itself using optical and electron microscopy. To measure the residual stresses after partial fatigue, X-ray diffraction measurements were made. The surface layers were etched away and the measurements repeated to give a depth profile.

Two different metals were used in the study; a medium carbon steel and a high-strength aluminium zinc alloy. At the (relatively high) stress levels used the shot peening extended the fatigue life of the steel specimens by a factor of about 2, and the aluminium alloy specimens by a factor of 5. The lower factor obtained in the steel was attributed partly to the high stresses applied, and partly to fading out of the residual stresses. When the steel specimens were re-peened they regained their original values of compressive stress, whereas the aluminium specimens were not affected by re-peening.

The shot-peened specimens were found to have different shapes of crack, tending towards an elliptical shape with the crack surface length smaller. There was more sub-surface crack growth in the aluminium specimens, which may account for the ineffectiveness of re-peening in the investigation.

8.9 Fracture toughness testing

This is quite different from impact tests; it uses specimens which simulate a partly failed component. This is important in assessing damage toleration in service. The idea of fracture toughness testing is that a fatigue crack is grown in the specimen, then the stress intensity factor is obtained from the specimen dimensions and the load required to extend the crack catastrophically. The stress intensity factor is related to the elastic stress field around the crack tip. After fatiguing, the crack must be long enough to cause failure before general material yield. When viewing the fracture faces after failure it is noted that during fatigue the crack shows striations, with each striation corresponding to a stress cycle, whilst for fast fracture the face shows ductile dimples or cleavage facets.

The rules for such testing are too lengthy to be included here, though it is noteworthy that in a flat specimen there is a three-dimensional (thickness) effect which at first sight seems unexpected. If a specimen is too thin it is proportionately less liable to crack growth; a property sometimes made use of by laminated construction. The reason for the thickness dependence relates to the transverse (Poisson) contraction. Naturally this would be greatest where the stress is greatest, at the crack tip. In a thin specimen this contraction is able to take place more freely than in a thick specimen; further from the surface the lower-stressed surroundings inhibit the contraction in the most highly stressed part and thus impose a third-axis tensile stress. The thickness dependence is most marked when the plastic zone size at the crack tip is comparable with the sheet thickness. When testing thin specimens it is recommended to keep the stresses tensile to avoid buckling.

Many materials show a temperature dependence, being more brittle at low temperatures. Pure ductile metals and some alloy steels have a very high fracture toughness, and polymers a low fracture toughness unless reinforced. Ceramics normally have a very low fracture toughness, making standard tensile testing rather difficult.

In applying fracture data to specific situations, safety factors of between 2 and 4 have been suggested. It is not clear at the moment whether these factors refer to acceptable stresses or acceptable crack depths.

Chapter 9

Surface Profile, Friction, and Wear Measurements

9.1 Surface profile measurements

The surfaces of manufactured components are not smooth but have a bumpy surface profile with hills and valleys. The surface profile can be measured with a surface profilometer, which is usually a stylus type, having a very fine stylus on a cantilever which is dragged a few millimetres across the surface. The vertical movement of the cantilever is amplified to produce the surface profile. Usually the profile is displayed with the vertical scale magnified much more than the horizontal scale, up to 1000 times. This gives the visual impression of a very spiky profile with almost vertical-sided peaks and valleys, rather like the Alps or the Himalayas, whereas the truth is that the surface is usually more like a gently undulating 'rolling' countryside, with slopes of less than 10%. This is illustrated in Fig. 9.1.

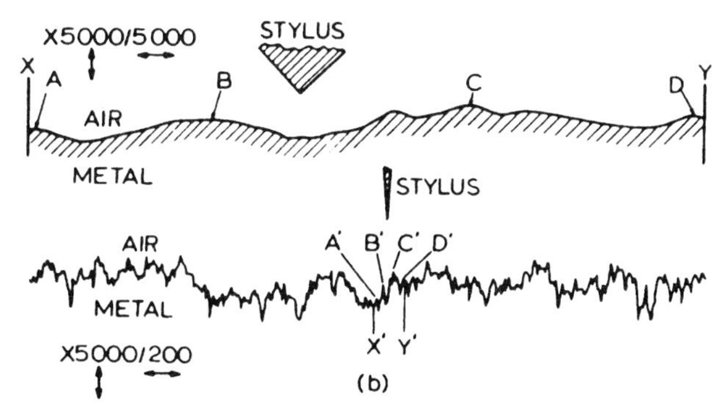

Fig. 9.1 Surface profile trace; effect of horizontal compression (from *Tribology Handbook*)

Surface profilometers are available in various forms ranging from bench-mounted instruments down to portable devices the size of a cigarette packet. The portable devices are very convenient for measurements *in situ* on machines or for large components which could not be brought to the laboratory. The disadvantage of the portable machines is that they usually only measure a single quantity, whereas the more refined bench instruments can perform more refined analyses. A very well known make of instrument is the 'Talysurf', and this name has become almost the generic name for surface profilometers.

The stylus-type instruments often have a skid which moves along the surface with the stylus and provides a (slightly varying) reference position as well as physical protection for the delicate stylus. When a skid is used, it in effect filters out most long-wavelength features, and thereby compensates for large departures and curvature of the surface, whereas if a skid is not used, the profilometer can record all deviations of long or short wavelength.

Most traditional stylus instruments incorporate simple electrical filtering which effectively attenuates features longer than a certain wavelength, known as the cut-off length or sampling length. Standard settings are usually used, of 0.25, 0.8 or 2.5 mm, to suit various typical surface finishes, and it is important to note which setting is used when recording or comparing results. Digital instruments can have more refined filtering methods, and it is believed that results can vary significantly depending on what type of filtering is employed.

There are also non-contacting three-dimensional (3-D) profilometers, using lasers to detect the surface heights. These are, at the time of writing, very expensive devices, but they have the merit that they can readily scan and analyse an area, not just a single line on the surface. Stylus instruments can also cover an area by multiple scans.

Various values can be calculated from the surface profile. A commonly used one is R_a (roughness average), also known as CLA (centre-line average). This is the average vertical deviation from a line drawn centrally through the profile, and it is a good general indication of smoothness. Machine turning gives R_a values typically between 5 and 0.5 μm, and grinding gives between 1.5 and 0.1 μm.

Note that techniques such as diamond (or CBN) turning can give R_a values as good as for grinding, but the shape of the profile may be substantially different. The R_a value therefore is clearly not a complete description of the nature of the surface, and there are various other parameters which can be calculated from the profile to provide a fuller description. However, for practical purposes, the R_a value in combination with a description of the manufacturing process, is usually an adequate definition.

Another commonly used value is the peak-to-peak height, usually expressed as R_z. This is obtained by averaging the five highest peaks and the five deepest valleys in the measured profile. Values of R_z are typically about five times R_a for turned or ground surfaces. Measuring instruments may be switchable between R_a and R_z, so care needs to be taken in using the instrument and recording results.

Other quantities, such as R_q, the RMS surface roughness, can be calculated by some instruments, or can be approximately estimated from R_a values. It should be noted that the values produced for any of the roughness parameters can be affected by the way in which the instrument works. Filtering of the output, and the choice of sampling interval, will affect results.

It should be noted that running-in can have a very beneficial effect on surface finish, often improving it by a factor of between 2 and 5. Running-in also changes the shape of the profile, by removing the tops of the peaks while leaving the valleys untouched. There are quantities which can be derived from the profile measurements which express this effect numerically, but these are beyond the scope of this text. The effect can be seen clearly if surface profile traces are taken before and after running-in.

For a machined (turned or ground) surface, the profile and roughness values can vary depending on whether the measurement is taken parallel to the machining direction or perpendicular to it. The most representative measurement would logically be in the direction in which relative sliding would take place in service. Unfortunately, for shaft journal surfaces this would mean measuring around the circumference, and most instruments are only suited to measuring in approximately straight lines. In practice, surface profile measurements are usually taken in the direction transverse to the machining or grinding direction.

The main significance of surface roughness is the effect it has on tribological components such as bearings, gears, etc. If the surface, particularly of the harder component of a mating pair, is too rough, this will result in an unduly high wear rate. For lubricated bearings, the roughness (after running-in) should be less than the expected oil film thickness to ensure successful operation.

There are a few cases where the surface should not be too smooth. An example is engine cylinder liners, where excessively smooth surfaces fail to retain lubricant and this results in scuffing of piston rings and cylinder surfaces. Another example is on components which are to be pressed together and secured by adhesives. In this case a reasonably rough surface is beneficial to avoid wiping off the adhesive during assembly and to provide a good 'key'.

9.2 Friction measurement

Friction measurement is not usually a problem if the coefficient of friction is fairly high, above about 0.1. However, low coefficients of friction in the range 0.01 to 0.001 occur in lubricated tests and with some specific dry rubbing materials (PTFE and similar), and in this situation it may be quite difficult to avoid interference from the applied load and to measure the friction accurately.

Many teaching laboratories contain a simple apparatus for observing the frictional drag of a journal bearing, shown schematically in Fig. 9.2 It is tempting to observe only the angle through which the freely swinging load carrier is deflected; this ignores the shift of the load line. The error involved in ignoring this may be judged by considering that the nominal radial clearance between journal and bearing could be 0.0015 of the radius and the coefficient of friction might be around as low as 0.002. The bearing acquires an eccentric attitude due to the nature of the hydrodynamic film, giving a lateral offset of the order of 20 to 40% of the bearing radial clearance. This would lead to a discrepancy between the apparent torque on the bearing and the torque on the shaft, underestimating the friction by up to 30%. Reversing the rotation shifts the bearing eccentricity and the tilt of the carrier to the other side; the error continues.

This problem was recognized in the early days of modern bearing studies and led to the adoption of dynamometer-mounted driving motors so that the torque on the shaft could be observed independently. Even this approach requires care, because friction and drag losses elsewhere in the apparatus can be of the same order of magnitude as the friction being measured, and can be difficult to isolate.

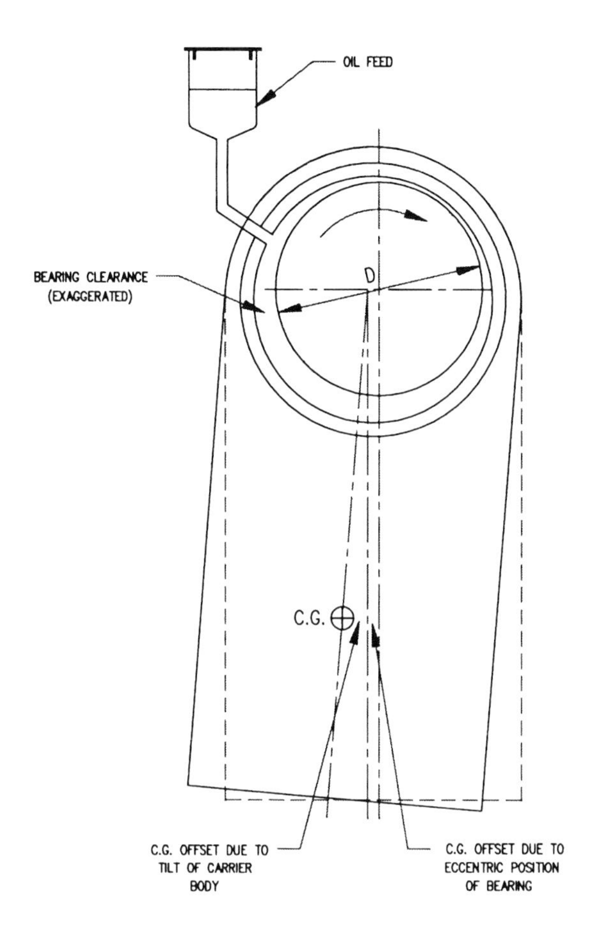

Fig. 9.2 Observing the frictional drag of a journal bearing

Detailed discussion of this topic is beyond the scope of this book. The main points are that low coefficients of friction are difficult to measure accurately, and simple procedures such as reversing the direction of movement may not help to detect the errors.

Fortunately, there are not many circumstances where it is necessary to measure small coefficients of friction accurately. The main use of friction measurement is in materials wear testing, where monitoring the friction gives a rapid indication of changes in wear regime or operating and environmental conditions.

9.3 Wear measurement

As wear is one of the main reasons for failure of machines, equipment, and other things such as human joints, wear testing is an important aspect of development of new materials and devices.

Wear measurement and testing is notoriously inconsistent, and it is very common for results to vary by at least an order of magnitude depending on the details of the test rig and the measurement method. Measurements have even been known to show negative wear, though not indefinitely!

There are various wear mechanisms which can occur, either individually or in combination. The principal ones are adhesive wear, abrasive wear, erosion, and surface fatigue. All these mechanisms can be affected by the operational and environmental conditions, for example, speed, surface pressure, temperature, presence of fluids and gases, to name but a few. There are a large variety of test rigs and test methods for simulating the various types of wear, and rather than describe the various types at length, it is appropriate to mention some of the pitfalls and problems. The discussion centres on the pin-on-disc test, but many of the comments apply equally to the whole range of test configurations.

Probably the most well-known test rig is the 'pin-on-disc' machine, where a small cylindrical specimen (the pin) is loaded against the face of a rotating disc. The load and the speed may be varied, and the working area can readily be enclosed and subjected to various gaseous or liquid environments. The wear can be measured either by directly monitoring the movement of the pin, or by periodically removing the pin to weigh it or measure it. The pin-on-disc and a selection of other common configurations are shown schematically in Fig. 9.3.

It is often desired to accelerate the wear rate in a rig test, as one may be trying to simulate a wear process which in service takes many months or years. Moderate acceleration is usually reasonably valid, providing that the material properties or the mode of wear are not unduly changed.

However, substantial acceleration is not always possible. For example, piston rings in service normally wear by a mild adhesive wear mechanism, the surfaces remain smooth, and only fine wear particles are generated. If the wear rate is accelerated in a test rig, then the wear mechanism may

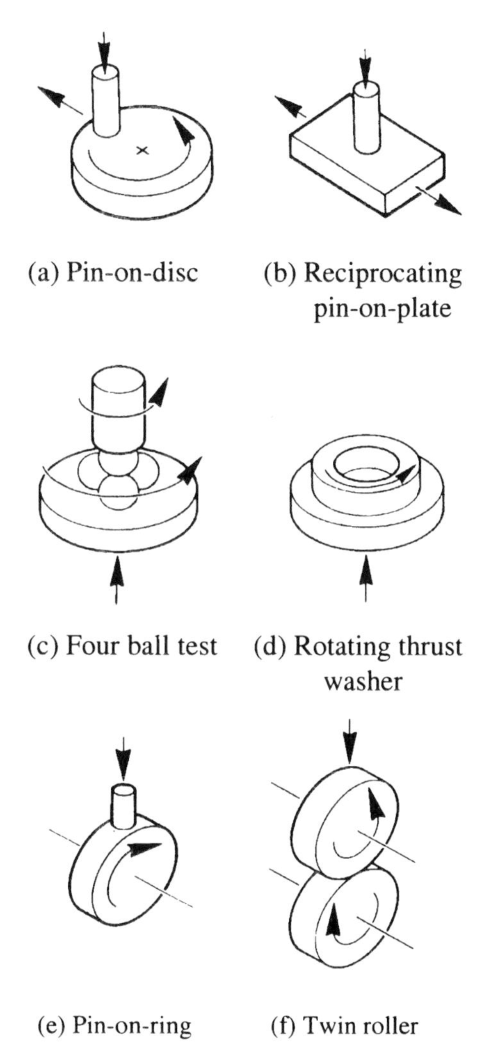

Fig. 9.3 Typical wear test rig configurations (reproduced by permission of Plint and Partners Limited)

change to scuffing, producing a dramatically higher specific wear rate, and surfaces resembling a ploughed field. This is because the material properties have been changed by the high frictional temperatures, and this has resulted in a changed mode of wear. Tests accelerated to this extent are usually invalid. Not only can the results not be scaled to an equivalent in-service situation, but one cannot even determine a valid order of merit for a range of materials tested. A material which performs well in service may perform relatively badly on an accelerated test, and vice versa.

As a general guide, accelerated tests can be considered as representative providing that the nature of the worn surfaces approximates to the in-service situation, and that the interface temperatures are not high enough to cause substantial changes in the mechanical properties of the materials.

Lubricants and surface films have a dramatic effect on friction and wear rates. For some materials, especially some ceramics and graphite, atmospheric humidity acts as a lubricant. For example, carbon/graphite alternator brushes were discovered to wear very rapidly on high altitude aircraft, due to the low humidity in the atmosphere. With plastics, the wear debris may form a surface film on the counterface, and any disruption of this film can dramatically increase wear rates. With metals, oxide films and microscopically thin adsorbed layers of hydrocarbons act as lubricants. Inadequate controls on these aspects can explain some of the major variations found with wear test results, even on seemingly identical experiments.

For example, variations in friction unrelated to the instrumentation used have been caused by the following:

1. During reciprocating friction and wear testing it was found that the solvent used to clean the specimens before use could significantly affect the results. Some of the solvents formed deposits on the surface which reduced the friction and wear in the materials.

2. In a series of tests conducted on dry (unlubricated) ceramics, it was found that at 10 a.m. every day the friction and wear rate of the specimens decreased significantly. After much investigation this was found to be the time when the operator opened his lunch box and peeled and ate an orange, the oil vapours from which could affect the results even when he was at a distance of a few metres from the test specimens.

3. A pin-on-disc experiment with ceramic on ceramic was conducted with water as a lubricant. After some time, the coefficient of friction rapidly dropped from around 0.1 to less than 0.01. The experimenter found that the surfaces had polished themselves to such a degree that a hydrodynamic water film a few Angstroms thick was developing and giving hydrodynamic lubrication.

Surface finish, particularly on the harder specimen (usually the disc in the case of pin-on-disc tests), can be important in determining the wear rate and the formation of surface films, etc. Even the direction of grinding or polishing on the disc surface may be important.

Wear rates usually vary with time. In the majority of cases there is a running-in period during which the wear is relatively high, followed by a long period of 'steady state' wear at a lower rate. The running-in process serves to remove some of the high spots, turning the peaks into plateaux, and therefore improving the surface finish and decreasing the effective contact pressure. This usually results in a reduced wear rate.

Sometimes, if the surfaces become too smooth, the wear rate is actually increased. This is because the valleys in a surface profile can act to retain lubricants, and if the valleys are smoothed out, the lubricant is lost, as in the piston ring and cylinder example mentioned above.

These effects need to be recognized and allowed for when carrying out wear testing and measurements.

Various methods are used for measuring the amount of wear, and these are discussed below.

Measurement of the movement of the pin towards the disc can be carried out continuously during a test. Unfortunately, the movement being measured is usually comparable with thermal expansion effects in the specimens and in the test rig as a whole. This method is therefore generally fairly insensitive, unless the thermal expansion effects are controlled (see Section 9.3.1), and is impractical for measurement of wear of thin coatings and surface treatments.

Weighing the specimen periodically is a commonly used technique. This can be insensitive if the wear to be measured is substantially less than the specimen weight. Other possible sources of error include weight reduction due to corrosion, or weight addition due to oxide layers, or absorbtion of moisture or lubricants. Also, it is difficult to ensure that the specimen is replaced in exactly the same position, so after each measurement there may be a period of higher wear rate as the newly positioned surfaces run in. Again, this method is not really suitable for measuring the wear of coatings and surface treatments.

By making the specimens non-conformal, e.g. ball-on-plate, or ball-on-ball, it is possible to deduce the depth of wear from a measurement of the diameter of the wear scar. This can be carried out without disturbing the position of the specimen, if the specimen arm can be swung to a position where the wear scar can be viewed with a microscope. This method has the advantage of being most sensitive for small amounts of wear, and is therefore very suitable for measuring the wear rates of thin coatings and surface treatments. Certain test methods (e.g. the 'ball cratering test') have been recently developed to exploit this measurement technique. In the ball cratering test, the ball is usually a standard steel ball from a ball bearing manufacturer, and the test specimen is a small plate with the appropriate coating or surface treatment.

Collection and measurement of the wear debris is a possibility, though not commonly used on test rigs. However, for wear measurements on machines in service this method is used extensively, providing there is a recirculating oil system or similar whereby the debris can be intercepted by a filter or magnetic plug. The technique is used for health monitoring of machinery, and also as a research tool. By doping a particular component with chemical tracers or radioactivity, it is possible to monitor the wear rate of, say, a single piston ring in a diesel engine.

Many wear test rigs incorporate measurement of the friction coefficient. This is usually straightforward and accurate, except in the case of lubricated tests where the friction coefficient may be very low (0.01 or less). Difficulties with these measurements are described in the preceding section.

The friction coefficient can be quite informative about the test operating conditions, as changes in friction indicate a change in wear severity, or a change in lubrication conditions. Therefore when setting up a wear test, an increase in friction may indicate that the loads or speeds have been increased too much and that a severe wear regime is being entered. An unexpectedly low friction coefficient could indicate incorrect cleaning of a specimen leaving a lubricating surface film. Measuring friction also acts as a cross-check on the test set-up. If the friction coefficient is substantially different from expectations, it suggests either that one or more sensors or instruments is indicating incorrectly, or that some test condition has changed.

Another method of monitoring conditions in a lubricated test is by measuring the contact electrical resistance. This is obviously only possible where both components are conducting. The contact resistance can give a good indication of the presence of a partial or full lubricant film. In cases with a full film substantially thicker than the surface finish of the materials, capacitance methods can be used to measure the film thickness.

Standard test rig configurations such as pin-on-disc have the merits of simplicity and economy of operation, and (in theory) relatively good repeatability. However, most such machines do not closely simulate the contact conditions in real machines and devices. Therefore pin-on-disc machines are often used for screening tests when evaluating candidate materials, coatings, lubricants, etc. for a particular situation. This then is best followed either by tests on the real machines, or by tests in purpose-made test rigs simulating more closely the real service conditions. As there are scale effects with some materials and also with lubrication mechanisms, it is best if these test rigs operate at or close to full-scale sizes, loads, speeds, etc., if they are to provide a reasonably close simulation.

Some materials, particularly resin/fibre composites, and also moulded plastics, exhibit different wear and friction properties in different sliding directions, and may have surface properties substantially different from the bulk properties. This is a further reason for preferring test rigs closely simulating in-service conditions where possible.

Manufacturers of bearing materials and other materials used in sliding and wearing applications often show wear rates in their catalogues and brochures. Perhaps not surprisingly, each manufacturer tends to run the tests under conditions which make their particular material perform favourably. Therefore, if using manufacturer's data, one should discover from the manufacturer what the test conditions were, and whether they are relevant for a particular application. Independent tests have been carried out on a wide range of polymeric and other bearing materials, under a wide range of loads and speeds, and these data are published by ESDU (Engineering Sciences Data Unit) in the form of a selection and design guide (**35**). This is well worth referring to before embarking on a series of tests, as even if the exact material being contemplated is not included, other broadly similar materials will be, and the data will give guidance on the results to expect.

It is important to note that scale and configuration effects can be very significant, so merely to reproduce the operating speeds and pressures may not be a sufficiently close simulation. One of the main reasons is dissipation of heat, which affects the surface temperatures at the wear interface. Most laboratory test rigs have quite good dissipation due to small test sample sizes and open construction, while the service situation which one is trying to simulate may have relatively poor dissipation. It is important to be aware of this, and where possible to measure or estimate the temperatures from the service situation, and if necessary apply artificial heating on the test rig.

Test rig configuration (e.g. pin-on-disc, block-on-ring, reciprocating pin-on-plate, etc.) can have a large influence on results, as can the orientation of the specimens. The orientation is particularly important in relation to the wear debris. An orientation which tends to trap or recirculate wear debris may accelerate wear if the debris is abrasive (e.g. iron or iron oxide), or decelerate wear if the debris is benign (e.g. PTFE). The general guidance is to choose a configuration and orientation which is, where possible, close to the service conditions which one is trying to simulate.

9.3.1 Problem examples: wear and friction test rigs

Loading arrangements

Using dead weight loading can lead to errors caused by resonance of the load arrangement giving unrepresentative severe friction and wear results. On one test rig, at high loads and speeds vibrations of the load arm were noted, and calculations showed the natural frequency of the loading arrangement fell within the operating range of the machine. This was solved by changing the aspect ratio of the load arm, making it narrower and deeper, without affecting the load at the specimen. Loading arrangements using springs or pneumatics are generally preferred for high loads, as these provide some resilience and damping and give more repeatable results. Conventional pneumatic cylinders exhibit friction stick–slip effects, and pneumatic bellows or diaphragms are best. For light loads, dead weights are convenient and accurate, providing that vibration is avoided.

When conducting reciprocating friction and wear testing, it is preferable to have the line of application of the force in line with the contact, to avoid the specimens attempting to rock at either end of the stroke, which can cause significantly increased wear.

Measuring wear rate on a pin-on-disc rig

On an industrial pin-on-disc rig, where high accuracy of wear measurement was required, a linearly variable differential transformer (LVDT) was used. This consists of a freely moving core, and coils for energizing and for pick-up of the signal (see Section 3.1). It had effectively infinite resolution, and a linearity within ±0.5%. It had a very low spring force; just sufficient to keep in contact with the specimen carrier, which therefore did not affect the applied loads. This worked extremely well as long as the displacement between the upper and lower specimen carriers was purely due to wear of the specimens.

As the specimens would heat up during the course of the test, or could be externally heated, thermal expansion of the specimens, specimen holders, and transducer mounting would have changed or even dominated the apparent wear measurement. To minimize this effect, very low thermal expansion materials were used for all components except the specimens themselves. The temperature of these was measured, and therefore the thermal expansion could be calculated.

Measuring wear rate on a reciprocating rig

On a high-speed reciprocating friction and wear test rig running at 35 Hz and a stroke of up to 50 mm, it was required to measure the wear using a non-contacting device. This was done using a capacitive type proximity sensor attached to the stationary specimen mounting, and a horizontal 'flag' which reciprocated with the moving specimen.

The flag had to be slightly longer than the 50 mm stroke, and problems were experienced in setting up and keeping the flag horizontal to give a constant output from the sensor. As the reciprocating specimen wore, the flag began to dip. This was avoided as far as possible by starting with the flag tilting slightly upwards at the beginning of the test, so that it would gradually wear itself horizontal, and then slightly downwards. The variation in signal due to the tilt could be compensated by appropriate signal processing, since only progressive wear rates rather than instantaneous values were required.

Errors in measuring traction force on a rolling/sliding rig

On a machine used to simulate rolling and sliding contact, e.g. of gear teeth, using pairs of test rollers, a strain gauge load transducer was used to measure the traction force at the contact. At a certain slip rate this was also picking up machine vibrations which completely swamped the traction signal. The error was observed by connecting up to an oscilloscope and observing a roughly sinusoidal signal, where a steady force had been expected. Elastomeric isolation of the transducer was tried, to attenuate the vibrations, but the flexibility of mounting then affected the position of the contact. The solution adopted was to apply appropriate electrical filtering of the signal from the transducer. This is a typical example of the difficulties associated with friction (in this case traction) measurement.

Problems with wear test specimens

A major test program was conducted on the life of PTFE-type coatings for use on frying pans, and different materials were being compared, when it was found there was an inconsistency in the results. There were two operators who applied the bonding layers, and it was found that when the layer had been applied by one of the operators, the coating would last more than twice as long as those applied by the other operator. The quality control on application of the bonding layer was a more significant variable than the type of coating material.

This demonstrates the problem that quite subtle differences can have large effects on wear test results. For this reason it is wise to do at least two or three tests under nominally identical conditions to check for consistency. Variations of +/– 30% are quite normal, but much larger variations suggest something is amiss. Inadvertent contamination from airborne vapours either before or during the test is a common cause of problems.

Chapter 10

Internal Combustion Engine Testing

10.1 Engine variability

Internal combustion (IC) engines suffer from considerable variability, not only from engine to engine, but from test to test. Even a basic quantity such as the peak output torque may vary by +/–2% from one test to another on the same engine under the same conditions, and variations from one engine to another can be worse, due to variations in build tolerances.

There are also variations which occur due to the test conditions, principally the combustion air conditions which are difficult (or expensive) to control completely. For this reason, there are internationally accepted calculation procedures to correct the engine output to that which would be expected at standard test conditions. These are documented in BS 5514 (**36**), and the equivalent ISO Standard. The principal corrections are for air temperature, air pressure, and humidity.

It is necessary to monitor the external air conditions on a regular basis during testing, as atmospheric pressure, temperature, and humidity can vary significantly over a period of a few hours. While continuous monitoring is not essential, readings every 2 hours are recommended for accurate work. Actual temperatures at the engine intake need to be monitored continuously, as these can fluctuate substantially over a much shorter timescale. If the engine air intake is not directly open to atmosphere, then the pressure difference between intake and atmosphere should be monitored, preferably continuously. With multiple engine test cells fed from a common air duct, there could be unexpected fluctuations due to the other engines.

A general recommendation with engine testing is to repeat the first test point at the end of a test sequence, and if any substantial variation is found, other than ones expected from changes in air conditions, etc., the engine requires checking for faults or deterioration (or improvement, such as running-in, or self clearing of deposits).

10.2 Torque measurement

Torque measurement is traditionally by means of a dynamometer which absorbs the engine power. The various types of dynamometer are not discussed here as they are well covered elsewhere [Plint and Martyr, (**37**)]. The reaction torque at the dynamometer is measured by means of a torque arm and a load-measuring device which may be weights, a spring balance, or a load cell. Quite commonly there are additional weights provided so that the indicator can be more sensitive. These additional weights, on the torque arm, must not be forgotten when calculating the torque.

There are two problems with torque measurement. Firstly, the dynamometer has to be suspended in pivot bearings, which may have friction which reduces the accuracy of the reading. There are ways of eliminating this friction, e.g. by using flexure pivots instead of bearings, or by having hydrostatic bearings. In practice the general vibration from the engine means that the dynamometer pivot friction is not usually a problem when running. Such measures are only needed for very accurate work, or for measurements where there is little vibration such as on turbines, or motoring friction tests on IC engines.

The second problem is that when measuring transient torques during acceleration or deceleration, the inertia of the dynamometer rotating parts affects the torque measured. This can be compensated by calculation if the acceleration rate is accurately known, but an alternative is to use an in-line torque transducer. These are usually strain-gauge-based devices mounted on the shaft connecting the engine to the dynamometer, and they transmit their signal by slip rings or by short-range radio or infra-red telemetry.

10.3 Air flow measurement

The pulsating nature of the air flow into an engine means that special precautions are required for measurement. Probably the most accurate method is to draw the air from a large airbox, measuring the flow rate by an orifice plate or venturi at the intake to the box, where the pulsations are somewhat smoothed out. Unfortunately, the typical box size required is of the order of several hundred times the engine capacity, which may be inconveniently large.

A frequently used method is the viscous flow meter (Fig. 6.5, Section 6.5.3). The air is drawn through a metal element at low speed, for example a corrugated, spirally wound strip which creates passages of around 1 mm width and perhaps 100 mm length. The length-to-width ratio ensures that the flow is essentially viscous rather than turbulent; hence the pressure drop varies approximately linearly with speed and the average pressure drop closely represents the average speed if the fluctuations are moderate. For extreme fluctuations, for example with a single cylinder engine, errors of the order of 2 to 3% are likely, as described in Section 6.5.3.

It should be noted that as this type of air flowmeter is not 100% linear, it requires calibration on a steady flow rig over the full range of flow rates expected in use. The meter requires protection from dust and deposit build-up in the small passages, usually by means of a paper air filter. Even with these precautions, periodic cleaning may be necessary to retain accuracy. The approaching air flow should be straight, free from swirl or bends. Serious discrepancies have been found when a bend was placed just upstream of the meter.

Performance of an engine may be strongly dependent on the configuration and detail dimensions of the air intake system, and the systems described above may interfere with this significantly. A small meter working on the corona discharge principle (e.g. the Lucas–Dawe meter) can be fitted into the intake ducting on an engine, therefore not disturbing the performance characteristics. This type of meter is good for transients, but sensitive to air temperature and humidity. Hot wire anemometers have also been used in the same way, but also suffer similar problems.

Engines with electronic fuel injection have air flowmeters as part of the injection system, so by calibrating this against an accurate reference at various engine operating conditions, the built-in air flowmeter can be used for non-invasive flow measurement. Many of these built-in meters are of a spring-loaded vane type, and their resolution (if not sufficiently high) can presumably be enhanced by fitting a shaft encoder.

10.4 Fuel flow measurement

Diesel and petrol (gasoline) fuels vary substantially in density, quite often there being a general shift from summer to winter grades of fuel sold from the same outlets. The calorific value per kilogramme is relatively constant, but the calorific value per litre can change substantially. Temperature changes also affect the density to a surprisingly large extent. The carbon–hydrogen ratio varies with the proportion of aromatic (ring-type) compounds coming into the refinery, and while this may not affect engine power output greatly, it will affect the gaseous emissions.

One way of avoiding part of the problem is to use a standard reference fuel, and to control its temperature at the point of measurement. This way, standard volumetric measuring techniques can be used without any complications. If normal commercial fuel is used, then density needs to be checked on every batch, and corrections for temperature applied.

Laboratory engine tests traditionally use a well-established form of fuel flow gauge shown in Fig. 10.1a. At first, steady state running is established with fuel coming from the high level tank, by gravity. Then the fuel tap T is turned off so that the fuel drawn by the engine comes from the measuring vessels, shown as the traditional globe pipettes, and is timed by a stop-watch. As soon as the measurement period is finished, the header tank is re-connected by opening the tap.

On one test the time for 50 cm^3 of fuel to be consumed became surprisingly long, indicating incredibly low specific fuel consumption. It was not until the engine started misfiring towards the end of a measuring period that the answer was found. There was a partial blockage between the globe system and the engine. The fuel supply was just adequate when running from the header tank. However, during measurement the head available to drive the fuel was lower, being just the head from the globes to the engine, hence the engine drew fuel not only from the globes but also an unmetered amount from the float chamber C, (Fig. 10.1b). After each measuring period the head from the tank was sufficient to refill the chamber. When the blockage was cleared the fuel consumption readings became credible.

This particular problem would not occur with a diesel or fuel injection gasoline engine, because in these cases there is no float-chamber or equivalent.

Gravimetric fuel measuring devices have been developed, which automatically compensate for density variations in the fuel. These are expensive but highly accurate. They also do not suffer from a problem with diesels, which is that the injector spillway flow, which must be returned to the measuring device, tends to contain air bubbles. With a volumetric measuring device, the bubbles inevitably upset the measurement, whereas with a gravimetric device, they do not.

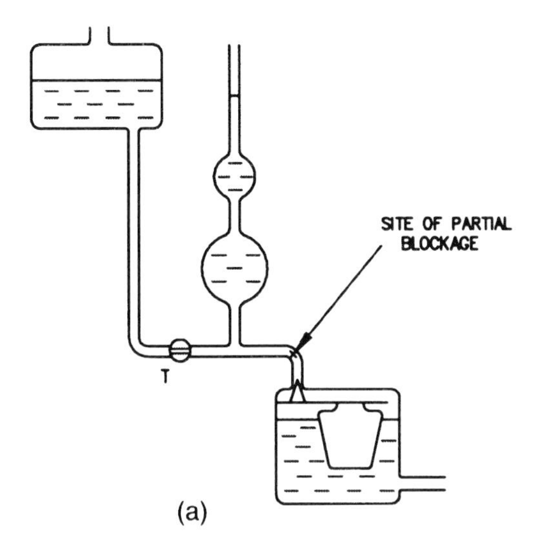

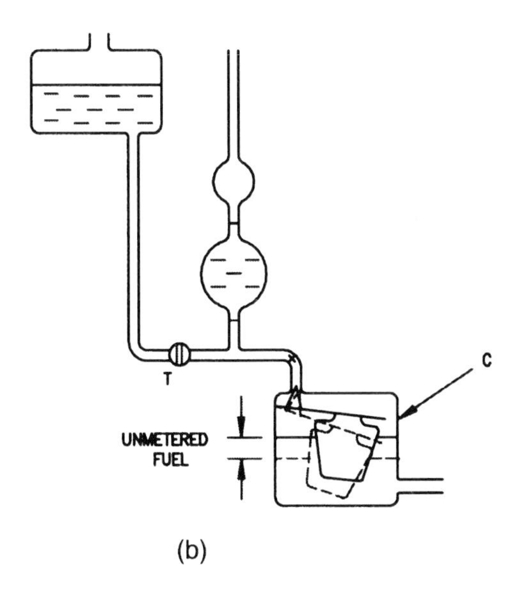

Fig. 10.1 An error in fuel flow measurement

For larger engines, coriolis effect flowmeters have been developed, which measure the coriolis forces induced by the flow of fuel through a vibrating U-shaped tube. These measure mass flow rate directly, and are not affected by small bubbles, unless they accumulate to create a large air pocket.

All continuous fuel flow rate measurement techniques are prone to errors, particularly at low flow rates, so where possible the measurements should be checked against a cumulative measurement.

10.5 Engine friction measurement

Engine friction is important as it has a large influence on fuel consumption of road vehicles, particularly at part load conditions, where the friction may correspond to one-third or more of the engine output. There are various ways of measuring friction, all with disadvantages and inaccuracies.

A motoring test, where the engine fuel and/or ignition are cut off suddenly, and the engine is driven by the dynamometer, appears at first sight to be an accurate method. If the readings are taken quickly, the engine is still at full operating temperature, so oil viscosity, etc. should be realistic. Unfortunately the cylinder pressures have reduced, so the piston ring friction is reduced. Also, the gas flow changes, so the pumping losses in the engine change. This method is therefore not highly accurate.

Morse tests are a variation on the same theme, where one or two cylinders are disabled at a time. Traditionally this was done by disconnecting one spark plug, but with fuel injection systems it is appropriate (and safer) to cut off the fuel to individual cylinders. Morse tests are also not particularly accurate, the reasons being the same as for motoring tests.

Willans line tests are based on the finding that for diesel engines, the fuel consumption versus load graph is very nearly linear over the lower part of the load spectrum. Therefore extrapolating this line back to zero fuel consumption gives a theoretical indication of engine friction. Unfortunately, this is friction at zero load, which is unrealistic for the same reasons as above.

The most appealing method is accurately to measure the cylinder pressure and crank angle, and to deduce from these the indicated mean effective pressure (IMEP). The friction could then be calculated from this and the brake torque. Systems for measuring the IMEP do exist, but have some limitations on accuracy, as described below.

Measuring the position of top dead centre (TDC) to sufficient accuracy is quite difficult, partly due to torsional deflections in the crankshaft when running, which can amount to something of the order of 1°. A 1° error in angle may give a significant error in IMEP.

Cylinder pressures are best measured with piezo transducers. If directly exposed to the combustion chamber, they often suffer from transient thermal effects during combustion. The radiated heat pulse causes a transient bowing of the diaphragm on the front of the transducer, affecting the pressure readings. This can be reduced but not eliminated by an insulating layer. If the transducer is recessed into a cavity, then resonance effects in the cavity may affect the readings. One approach has been to incorporate the transducer into a spark plug washer (or diesel injector washer), which presumably eliminates some of the above effects, but may make the reading sensitive to vibration and resonance effects.

Cylinder pressures often vary significantly between cylinders, so for a multi-cylinder engine it would be necessary to measure on all cylinders simultaneously.

While engine friction measuring methods may not be strictly accurate, they are nevertheless very valuable experimental tools for developing improved engine designs.

10.6 Specialized techniques

Emissions measurement is a specialized set of techniques, beyond the scope of this book. The automotive industry requires complex tests, based on various driving cycles, with closely controlled conditions. Many of the worst engine emissions occur during transient conditions, which is why these requirements have evolved. Further details can be found in Plint and Martyr (**37**).

Temperatures of the internals of engines are measured routinely as part of engine development. For static components, thermocouples are the standard technique, but for moving components, other techniques are used.

For pistons, it is common practice to fit 'Templugs', which are small slugs of low-melting-point alloys pressed into small holes drilled in the piston. A variety of plugs which melt at different temperatures are fitted, and the engine is run then dismantled to discover which have melted. It is important to decrease the load and speed of the test progressively, as heat soakage can result in parts of the piston becoming unrealistically hot if the engine is suddenly stopped from full power.

An arrangement has been described in Section 5.5 for measuring piston and piston ring temperatures on a research engine using an infra-red non-contact method. A small window is fitted in the cylinder wall and the infra-red detection instrument views through a stroboscopic disc. By varying the phasing between engine and stroboscope various parts of the piston can be observed.

Measurement of wear of piston rings has been carried out by irradiating a ring, then subsequently measuring the radioactivity of the engine oil. This technique could be applied to any component, and can be potentially used in the field as well as on the test-bed. Chemical doping rather than irradiation may be equally appropriate.

Diesel fuel injector needle lift can be measured, with a tiny magnet attached to the back of the needle, and a Hall effect sensor fitted into the injector housing. This is very sensitive to the position of the magnet with respect to the sensor, and has to be mounted very close to give a good result. The Hall effect sensor is probably the only device small enough and rugged enough to work in this application.

Chapter 11

Computers

The advent of computers has revolutionized nearly all branches of engineering. They are widely used in design, analysis, manufacture, process and industrial control, data collection and processing, and numerous other areas. In the field of engineering measurements, personal computer (PC) based or mainframe computers or purpose-built hardware are used in feedback control of processes and in data acquisition. Digital signal processing and intelligent instrumentation for error corrections are also used. Test equipment is often computer controlled, which allows for a great deal of flexibility in test conditions, simplifies data acquisition, display of test variables and feedback control of test parameters, and allows for multi-stage tests to run unattended and shut down automatically under chosen conditions.

A lot of experimental work is carried out by modelling the desired set-up on the computer and simulating the experimental conditions, for example stresses and strains, heat and fluid flows. This is normally much cheaper and quicker than conducting full-scale experiments, so that a number of different set-ups can be tried. However, it is preferable to confirm results of the simulation with physical measurements wherever possible. The physical situation may be limited, for example by the number of strain gauges that can be fitted or the number of probes inserted, and the simulation can help target critical areas. The measuring process can also affect the results, for example inserting thermocouples into the cylinder in IC engine testing can disturb the gas flow.

11.1 Control and data acquisition

Computers are nowadays commonly used for control of equipment, data acquisition, and analysis of the data obtained. Control and data acquisition systems are used for industrial and process control or with test equipment. In process control packages a virtual user interface or 'mimic' diagram may be used, which looks like a control panel with push buttons, warning lights, trending graphs, etc., but is operated with a mouse. As well as purpose-built hardware, a number of PC-based off-the-shelf control and data acquisition software packages are available, which then have to be set up by the user for the particular equipment being used. Other packages are sold with the equipment and tailored to operate it.

11.1.1 Off-the-shelf packages and some of their limitations

At the time of writing, there are at least four major, widely used, off-the-shelf control and data acquisition packages available in the UK. Of these, Labview is very popular; a Windows-based package by National Instruments. This is fairly straightforward to operate, being very graphical. The set-up consists of blocks of parameters, for example analogue inputs, connected on the screen by wires drawn by the user. Other major packages include Testpoint, sold by Keithley Instruments, HPV by Hewlett Packard, and Labtech Notebook.

Although the packages are relatively easy to set up for simple applications, they become unwieldy for complex applications and really require a trained programmer to set them up. This somewhat defeats the object of an off-the-shelf package. They can also be slower at executing programs than purpose-built software, particularly where there are a lot of calculations to be done, or when operating with large arrays. Most packages have maths functions built in, and additional packages, e.g. statistical analysis, available as add-ons. It is also easy to transfer data to spreadsheets for more complex analysis.

The data acquisition hardware normally consists of plug-in cards, with software drivers to operate it. For most applications the card will have a single analogue-to-digital (A/D) converter and the data are measured sequentially, with an input multiplexer switching between the channels. Most data collection is done at rates of around 10 to 100 channels per second, whereas in wear testing, for example, measurements may only require datalogging every couple of hours.

A typical fast data acquisition system could have 16 channels logging data at 100 kHz. However, for applications such as engine testing it may be necessary to record pressures and temperatures at intervals of 1 ms or less, and if measuring three or four pressures in the cylinder, to record them simultaneously. This requires either separate A/D converters for each channel or a simultaneous sample-and-hold facility. Boards are available to record data into the GHz range, as used in radar measurements and ultrasonic 'time of flight', when recording the time taken for a reflected signal to return to its starting point. One-off events such as impact tests can require extremely high data acquisition rates at the point of impact. Data acquisition rates are limited mainly by cost and the time it takes to convert from analogue to digital signals

One mistake is to select too high a data acquisition rate for the parameters being measured, and end up with too much data to store sensibly or interpret. There is also little point in collecting data at a rate higher than the response time of the instrument used to measure it.

It is sometimes useful to record data at a high rate at the start of a process if parameters are changing rapidly, then sample at a much lower rate thereafter, or collect the data only for a single typical firing cycle, for example. Also during engine testing it would be useful to record the bulk of the data around the top dead centre (TDC) position when all the action takes place, then say every 5° for the rest of the cycle. However, this is not always possible, as the A/D converter may be set for a fixed recording rate. It would also necessitate that the data be time stamped, which would mean collecting twice as much data.

Normally in engine testing the data collection is triggered by a shaft encoder indicating the position relative to TDC, and the data are position rather than time stamped. This means that the output is independent of engine speed. Whilst some of the parameters such as pressure need to be recorded at a high rate, others such as coolant temperature change only very slowly.

Most of the off-the-shelf packages are based on Microsoft 'Windows'. However, Windows-based applications are not really suitable for industrial and process control. Firstly, the system can crash, with results varying from inconvenient to disastrous. For one-off tests, for example exploding an aeroplane where losing the data would be disastrous, it is recommended to record the data on two systems simultaneously, or use a non-Windows-based system. Windows NT is more stable than the other versions, and various add-ons are available to try and get round the problem.

Secondly, Windows is not time deterministic, so that the response times of the system are not predictable. DOS based systems are more reliable and time deterministic. However, emergency stops and interlocks should always override the computer for safety, and in order to comply with health and safety regulations. The time aspect may not be important for most applications, but below about 10 ms Windows can be considered unreliable, and for high-speed feedback control, e.g. to operate a hydraulic piston, it is better to use some kind of hardware control. A lot of off-the-shelf hardware is available for motion control, for example high-speed position control of servo-motors.

11.1.2 Purpose-built computer control systems

In many cases purpose-built digital motion control systems are used in production lines, replacing the dedicated analogue systems used earlier. Although the capital expenditure for these may be higher, it is often possible to use a simpler mechanical design, and use the feedback control arrangement to increase speed of production, accuracy and repeatability, and to improve the adaptability of the system. The simpler mechanical arrangement can keep the running expenses low by improved availability in terms of less downtime, maintainability, and flexibility in terms of fast changeovers between product types.

In one example, a dedicated computer control system was used in the manufacture of chocolate-covered caramel bars. Both the quantity of product and the relative timing were critical parameters. The chocolate and caramel were very viscous at the machine operating temperatures, and hydraulic pistons were used to dispense the ingredients. The system consisted of a hydraulic servo actuator, with a stepper motor driving a rotary valve which activated the piston. This was driven by dedicated hardware, with an encoder for feedback control in case the stepper motor slipped. The piston travel was then proportional to the rotation of the stepper motor, giving good control over product delivery, and making it easy to change to different product sizes.

Another purpose-built microprocessor-based control system was used in a flow wrapping machine. Chocolate bars were to be wrapped at the rate of 600 per minute. With 185 mm length of wrapper for each chocolate bar, this meant that the paper and chocolate were both travelling at nearly 2 m/s. The system had to control the rate of product delivery from the main production line down the wrapping leg, including calling for more product from upstream, or slowing down or even stopping the crimping process to maintain the product queue. The product had to be positioned correctly relative to the paper; photocells would determine the position of each chocolate bar, then a variable speed conveyor driven by servo-motors would rapidly adjust its speed to give the correct product flow. With closed-loop control on the servo-motors, the loop closure time was around 2 ms. Each bar was monitored into the paper, and oversized bars could be thrown out before wrapping, to avoid jamming the machine. The paper was then wrapped around the chocolate bar, seamed along its length, sealed and cut off at the ends.

The new system allowed for much higher speeds of production than the old, where the chocolate bars were lined up nose to tail on a conveyor belt, slipping relative to the belt, and product pressure could build up from behind. Also there was much greater adaptability as the machine could compensate for oversize or undersize products and incorporate emergency strategies for when problems arose. The changeover between different products was by menu selection on a keyboard, and could be done simply by the operator in a few minutes, instead of by trained technicians over several hours. The simpler mechanical system made cleaning and maintenance easier, and the only problems encountered were as per the previous machines; build up of product on the drives, leading to problems with tracking or slippage, and chocolate on the photocells rendering them inoperative.

11.1.3 Types of control, and control of test equipment

A control and data acquisition package set up for use with test equipment will apply a series of test conditions, and then monitor the results. For example, test parameters such as speed or torque set points can be inserted into the package, the time for which each stage of the test should run, or the information that the test should move on to the next stage once a certain set point (e.g. temperature) has been reached. Both digital signals, such as signals to relays or solenoids, and analogue signals, e.g. for speed control, can be output.

The package will transmit the signals to operate the equipment, controlling the parameters perhaps on open-loop, closed-loop proportional or proportional-integral-derivative (PID) control, or even fuzzy logic control. Fuzzy logic consists of a number of rules, e.g. if the temperature is low, switch on the heater, and if a number of rules apply to the particular condition the controller gives them a weighting in order to calculate an output.

Open-loop control is not always a good idea. If there is a mechanical failure, for example the equipment has jammed, the control system has no way of knowing and the results can be catastrophic. Less noticeable errors can also arise. A CNC lathe was controlled by open loop, with a stepper motor used to drive a leadscrew for linear positioning. This was equipped with limit switches accurate to within 1 μm, and the zero was reset every time the head went back and hit the limit switch. However, as the screw warmed up during use it underwent a thermal expansion of considerably more than the requisite 1 μm, and the control system had no way of telling this.

Proportional control gives an output signal proportional to the error, and has the disadvantage that the parameter being controlled can never quite reach its given set-point. PID control overcomes this problem by increasing the output signal with time for a steady state error, but it must be tuned correctly for the step or ramp changes being used, so that it does not overshoot, oscillate, or take too long to reach the set-point.

An overshoot on temperature when testing lubricants, for example, could affect the lubricant structure or even set it alight. In one case a temperature feedback channel was set up incorrectly, and the PID controller commanded the heaters to continue heating even though the desired temperature had been exceeded. The oil being heated went over its flash temperature and caught fire. An automatic shut down if the temperature exceeds a certain value can be useful. An overshoot on load during, say, wear testing of materials could not only damage sensitive load cells but it could affect the surface microstructure or lubricant film and hence the results. A recommended maximum step or ramp rate for which the PID controllers are tuned may be specified with the equipment.

In Formula One motor racing, critical components on the cars are instrumented and the data from the instruments logged while the cars are on the circuit. This allows, for example, the wheel support components to be strain gauged so that the exact strains in service are known. Equivalent components are then mounted rigidly and actuators are used to apply loads equivalent to the wheel loading. Because the exact strains in service have been recorded, these are fed directly to the controller and reproduced under test conditions.

Setting up of the test itself should be done with care. A tractor dynamometer was supplied to a customer; this consisted of a tractor with d.c. dynamometers attached to each of the four axles to simulate the wheel. With much care and a certain amount of trudging up and down ploughed fields following a tractor, it was possible to set up the road load models so that a person in the driving seat said that it felt like the real thing.

However, an error in the control set-up meant that the ground velocity of the four wheels, which was a function of the angular velocity and the relative slip, could be substantially different whilst the tractor was theoretically travelling in a straight line. This gave the unrealistic condition of either the ground or the tractor stretching. Another minor problem experienced under test was that when running a certain part of the test code, the tractor would suddenly shudder to a halt. This was found to be a condition used when testing the software, in which the ground sloped up at 90°, in which case the tractive effort required should equal the weight of the tractor. It had been intended to remove this before testing on the actual equipment; failure to do so meant that the tractor had suddenly found itself attempting to climb up a vertical slope.

Computer control of testing makes it possible to set alarms on any of the measured parameters. For example, if the temperature were to get too high, flashing lights or alarm signals could sound (although these wouldn't be much help if the test was running unattended), and ultimately the machine could be shut down automatically before any damage occurs. When running tests unattended it is very important to pre-set the machine to shut down in the event of over-torque, over-temperature, etc.

There is, of course, a problem if the computer itself hangs up during a test. In one case cartridge heaters were switched on when the computer hung up, remained on, and the machine overheated. Either a watchdog function to check that the computer is still operative or a safety device such as a thermostat operating independently of the computer would be advisable in such a case. Any unattended test set-up which presents a potential hazard to equipment or people should ideally have automatic safety cut-outs which are independent of the computer.

11.1.4 Data acquisition and storage

Using computer data acquisition has the advantage that the data are already in a usable form, and can be input directly into spreadsheets for manipulation and graphical presentation without any laborious typing in. Trending graphs can be displayed on the computer while a test is in progress, instead of connecting the output signals to a chart recorder. This allows for automatic re-scaling of the axes during the test, and the number of parameters displayed is no longer limited by the number of pens.

The software may need to communicate with other equipment, e.g. using serial or networked communications. This means that there can be remote monitoring, whilst the control and data acquisition is done locally. In one case tests were being run on equipment in the UK, and monitored at the head office in the USA via the Internet.

The software may interface with a database in order to store data. This is useful in applications such as the military, where detailed recording of results is frequently required, or when complying with the ISO 9000 quality standards, where manufacture of each component must be traceable. It can also be useful in checking components during manufacture. For example when manufacturing components with a CNC lathe, it was possible to identify from the trend on increasing diameter when the tool wear was causing the component to approach its tolerance limits, and to change the tool before any components had to be rejected.

Calculations on the data can also be made in real time. In friction and wear testing, the friction coefficient is frequently recorded while the test is in progress, and the machine can then be shut down if the friction coefficient exceeds a certain value. This avoids the surfaces becoming so severely damaged that it is no longer possible to determine the wear mechanism in progress, or even the machine itself becoming overheated or damaged.

When a number of tests are to be run, it is important that the data are stored sensibly in a data file, with all important information related to the test conditions, ambient conditions if relevant, components used, etc., attached to it. The data files must be well labelled and easy to trace, and good housekeeping is essential as a lot of space can be occupied by the data.

11.1.5 Calibration, and limitations of a digital system

As for any test, for computer acquisition it is important that all the sensors being monitored are correctly calibrated, along with all downstream signal processing, and giving correct results in the correct units at the computer. In some cases, the sensor itself has a much higher capacity than the range being measured, for example on a dynamometer measuring motor torque a transducer with a 50 N range was selected, when the expected signal level was only 10 N, to allow for the kick on start-up. It is then advisable that the full-scale output of the transducer should match the expected signal levels, rather than the whole range of the transducer, so as not to give too coarse digitization at the lower end of the measured range. This will, of course, not help much if the resolution or repeatability of the signal itself is a limitation.

Converting from a digital to analogue output, for example when controlling motor speed, consists of changing from a series of steps to a linear signal by smoothing. Rounding errors can be a problem. When converting to a digital signal from an analogue input the linear signal has to become a series of steps, so it can be seen that if the steps are too coarse or the information sampled at too low a rate then information can be lost. Nyquist proposed that the sampling rate should be at least two and a half times the rate of the highest frequency component that is to be seen. If it is desirable to lose the high-frequency signals then a low-pass filter can be used.

Occasionally sampling at too low a frequency can give completely misleading results, with the apparent output signal looking completely different from the input. Aliasing is a problem which can occur when digitally sampling data at too low a rate. Figure 11.1 shows a sine wave being sampled at intervals of 10°; the output then shows up roughly as a sine wave of the correct frequency. Figure 11.2 shows the same sine wave

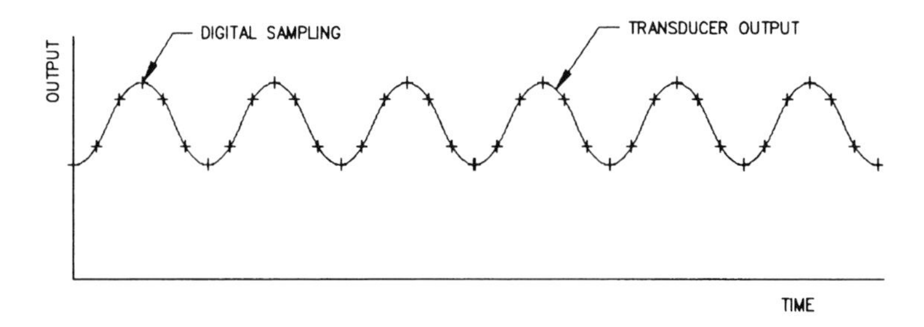

Fig. 11.1 Digital sampling at 30° intervals

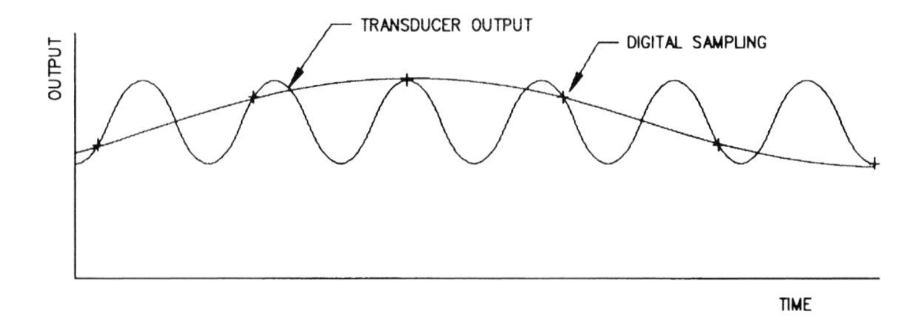

Fig. 11.2 Aliasing digital sampling at 390° intervals giving a long sine wave output

being sampled at intervals of 390°. This also gives a sine wave, but a very long wavelength one in this case, with a frequency completely different from the original signal. Aliasing produces output frequencies which are not real, and which are lower than the actual frequency being sampled. The output frequency due to aliasing is actually the difference between the original frequency and the sampling frequency.

11.2 Computer simulation

Computers not only act as a tool in collecting, processing, and presenting data, but also as a valid experimental device. Obviously a computer can only use physical data put in by the programmer (this includes ready-made software which may contain hidden assumptions), yet a computer simulation may give results which are not only quicker than a limited physical set-up but may also be more truthful, since a physical set-up is often a small-scale model and can suffer from unrealistic effects of heat loss, leakage, friction, etc. An example of this is mentioned below for illustration.

James Watt, when he met difficulties with a scale model of a Newcomen engine which he was overhauling for Glasgow University, became aware of the problems inherent in small-scale work. He found that the model worked perfectly but only for a few strokes before running short of steam. The small size exaggerated the energy loss caused by the steam condensing too quickly, which made him realize that there was no defect in the engine but that the defect lay in the scaling-down principle, there being no way of scaling the properties of steam. This realization led Watt to devise and develop his various improvements which led to a dramatic increase in steam engine efficiency.

It must be emphasized that a computer simulation is limited by the assumptions and simplifications which inevitably occur, and therefore cannot ever be a complete substitute for physical experiments. The benefit of the computer simulation is that it can quickly and cheaply try 'experiments' which can then identify promising avenues and eliminate dead-ends before expensive physical tests are undertaken.

Other examples follow, in which computer simulation allowed measurements to be made which would not have been practical or feasible otherwise.

11.2.1 Example: Computer simulation of an optical system

The case study in Chapter 3 on position sensing of photo masks refers to the development of a system for detecting contaminant particles on photo masks by shining a laser at the surface and interpreting the scattered image obtained on a CCD camera. It highlights the difficulties in differentiating between particles and the pattern of chrome on glass, which diffracts the laser beam at its edges.

Part of the project involved developing an optical system which would translate the image along the optical axis, and present a 'spatial filter plane' within the lens system. This gave a Fourier transform of the image, in which the diffracted light from the pattern edges formed a narrow line which could be removed by blocking with a strip of metal, whilst the scattered light from the particles would substantially bypass this. A low wavefront distortion was needed at this position.

To design the system a standard lens design package was used. Initially a design was input, which included the number and position of lenses, curvature, and separation of surfaces and glass types. The package then performed a 'ray trace', which takes selected rays from the object and calculates the image position, and works out how good the image is going to be. It obtained numerical values for the various aberrations or imperfections in ray bending which could arise; spherical aberrations (variation of magnification with radius), field curvature and distortion (differences in longitudinal and lateral magnification with field), and astigmatism and coma (causing differences in behaviour of light in different planes).

The design was then iterated to minimize the aberrations. In this case, to obtain a 1:1 magnification a symmetrical design could be used, which avoided both distortion and coma. Parameters to be varied included the power and position of each of the lenses in the system, and the shape and thickness of each lens, including the curvature of both surfaces.

To reduce the spherical aberrations, astigmatism, and field curvature, negative ray bending was needed, so a lens was split into positive and negative parts, and to keep the same magnification the negative lens was made from a lower refractive index of glass than the positive lens. To achieve the desired quality using catalogue lenses, the positive lens was first split into two, and later three, to give more degrees of freedom in altering the parameters. Aberrations were minimized individually for the lenses by altering their shapes, and cumulatively for the system by distributing ray bending evenly between lenses and using the computer to optimize the lens separation.

When a design of acceptable quality had been obtained, the computer established tolerances by introducing errors such as variations in power, thickness, centring, tilt, and so on, then calculating the resulting aberrations to see what was acceptable.

The use of the lens design package made it possible to experiment with the different designs and configurations until an optimum design had been achieved. The calculations used to be conducted laboriously by hand using teams of people, limiting the number of set-ups which could be tried. To conduct these experiments with actual lenses would have been too costly and time-consuming.

11.2.2 Example: Simulating release of fission gases in nuclear fuel rods

Another example where computers were used for simulation purposes was to investigate changes in the pore morphology of sintered materials (**38**), due to diffusion in the grain boundaries under internal and external pressure. This had particular relevance to nuclear fuel rods made of sintered uranium dioxide, and the release of fission gases, either during use, or when storing spent fuel rods. It was preferred initially to study these effects in theory rather than experimentally!

The gaseous fission products are insoluble in the uranium dioxide crystal lattice, and diffuse to the grain boundaries where they precipitate into bubbles. These grow and form connected pores through which the gases can be released, which may collapse again after venting. The process causes expansion of the fuel rods, helping to keep them in close contact with the cladding, but if excessive can damage it.

In order to predict the behaviour a set of differential equations were set up, firstly to balance the surface tension and the fission gas pressure to give an equilibrium configuration. Then equations were added to simulate the effects of surface diffusion and grain boundary diffusion, thereby giving a mechanism for the pore growth and collapse. The equations were solved numerically by computer, being too complex for hand calculations. Results were compared with measured values of porosity, using scanning electron microscopy for closed pore configurations, then extrapolating for the open pore case.

11.2.3 Modelling using structural finite element analysis

Computers are often used for modelling an actual situation, for example in finite element (FE) work. Structural FE analysis is used successfully in predicting stress concentrations around cracks and stress raisers, which can then be checked using strain gauge and photoelastic methods. For particularly critical applications, physical fatigue tests on the components may also be carried out. The benefit of the FE analysis is that it can be used to optimize the shape of the component at the design stage before making samples for testing. If sufficient is known about the performance of similar shaped components, then testing or stain gauge measurements may not be necessary.

Whilst the equations used in a FE analysis are relatively straightforward, it is the amount of calculation and iteration required to arrive at the solution which makes the use of computers so valuable. Often when using FE analysis to model a relatively straightforward situation, the results can be checked by hand calculations and/or physical measurements. Providing that the model is found to represent reality, it may then be extended to simulate a more complex situation which would be difficult to make and measure.

The analysis consists of taking a model of a structure, dividing it up into discrete 2-D or 3-D elements then specifying that each element should not only be in equilibrium but should also be compatible with its neighbours and behave in accordance with its material properties. The user sets up the model, defines the boundary conditions; loads or deflections applied, then leaves the computer to iterate until the forces or stresses and deflections for each element are determined. The mathematics assumes linear elastic behaviour, so cannot be readily applied to plastic deformations or other non-linear behaviour. However, non-linear situations can be modelled approximately by iterating the process, locally adjusting material parameters according to the state of stress and strain. Some FE packages include procedures for modelling plastic behaviour by this means.

The method has particular advantages for indeterminate structures, for example braced frameworks which could be extremely time consuming to evaluate manually. The areas of high tensile and compressive stresses should be highlighted, but regions in danger of buckling may not all be automatically detected, since buckling modes are an example of non-linear behaviour. Some FE packages contain routines to check for buckling, but it is not clear whether these are comprehensive. It is particularly important to check on long slender elements in compression, and thin-walled structures. Cardboard or sheet plastic models can also be useful to visualize and demonstrate deformations or buckling modes, particularly lateral effects in thin walls.

When using FE analysis it is advised to check the results with a rough hand calculation to ensure they are reasonable. People sometimes have more faith in a computer output than a manual calculation though both are subject to human error. It is only too easy to use inconsistent units, for example when inputting the materials data. A danger of computing is simply the speed with which results can be obtained; in slow manual calculations one spends much more time on a task, leaving better chances for lingering doubts to become conscious.

Other problems experienced include elements which have distorted beyond credible limits under high stresses, probably needing a finer mesh or different element type. NAFEMS (**39**), gives an example of a four-noded element under beam bending, where the element shows an unrealistic level of shear deformation. Here the stresses obtained by the analysis were only 20% of the actual stress levels, which could be dangerous. This would have been picked up by checking overall equilibrium of the structure, or discontinuities in stress at the interfaces. An eight-noded element gave a much better result, allowing more realistic distortions.

Finite element models can suffer from too coarse a mesh, so that areas of high stresses may not show up, or even too fine a mesh, which occupies too much computer time and memory, and may fail to converge. Incorrect element combinations can cause a problem, when one type of element is incompatible with another. This may occur when combining 2-D and 3-D elements, or buckling and other elements.

Inserting the correct boundary conditions in the model is important. For example, modelling as rigid a boundary which in reality is slightly flexible (e.g. a pinned joint) can give artificially high stresses at the boundary. This may not matter if it is not close to a critical region, but if in doubt the model should be adjusted.

Highly localized stresses above the yield point of the material flagged up by the FE analysis may in practice lead to local yielding of the material and a redistribution of the strains and stresses. If the loading is static (i.e. fairly steady and unidirectional), the material is ductile, and the region of yielding is small, then it may safely be ignored. However, if as is commonly the case, the loading is varying in magnitude and direction, then localized stresses of this magnitude suggest that either the model is inaccurate or the design of the component requires modification.

11.2.4 Example: A simulation of a pneumatic dashpot

The following points arose during a computer simulation of a pneumatic dashpot for a vehicle suspension. In the dashpot, air was displaced up or down through an orifice and it was necessary to keep count of the amount and pressure of air above and below the piston. The flow through the damping orifice varied as the square root of the pressure difference between top and bottom, which could change from upward to downward, i.e. positive to negative. When the air flow during a time interval was calculated, the computer stopped as soon as it was asked for the square root of a negative pressure difference. This should have been foreseen but caused a delay. It was easily overcome by preserving the + or – sign, then square-rooting the modulus (positive value) of the pressure difference.

Another fault due to inexperience was to ask for a division by a near-zero number because the displacement from static level was used in the denominator (below the line) in one of the equations. At a near-zero value the computer stopped because it obtained an excessively high answer. This can usually be overcome by re-forming the equation, or checking for near-zero values at each step before doing the calculation.

Simulations such as this are easily implemented using spreadsheets. This approach means that no programming knowledge is required, graphical output and further analysis and processing is easy. The problems mentioned above can be handled by introducing extra columns, such as a column which calculates the modulus of the pressure difference, and the division by a near-zero number would show up in the spreadsheet as an overflow.

References

(1) **Boyle, T. J.** (1973) Hope for the technological solution, *Nature*, **245**, 127.

(2) BS 7854: Part 1: 1996 (replaces BS 4675) *Mechanical vibration – evaluation of machine vibration by measurements on non-rotating parts; general guidelines* (British Standards Institution, London, UK). Equivalent to ISO standard, ISO 10816-1: 1995.

(3) **Kaye and Laby** *Tables of Physical and Chemical Constants* (Longmans).

(4) Catalogue: An introduction to micro-measurements (Measurements Group UK Limited, Stroudley Road, Basingstoke, Hampshire, RG24 0FW, UK).

(5) **Turner, J. D.** and **Pretlove, A. J.** (1991) *Acoustics for Engineers* (Macmillan).

(6) **Kail, R.** and **Mahr, W.** (1984) Piezoelectric measuring instruments and their applications, *Messen und Prufen*, **20**, 7–12.

(7) BS 1041: Parts 2 to 7 *Temperature measurement* (British Standards Institution, London, UK).

(8) BS EN 60584-1: 1996 (replaces BS 4937) *Thermocouple Reference Tables* (British Standards Institution, London, UK).

(9) Farnell Components Catalogue (Farnell Electronic Components Limited, Canal Road, Leeds, LS12 2TU, UK).

(10) RS Electronic and Electrical Components Catalogue (RS Components Limited, P O Box 99, Corby, Northamptonshire, NN17 9RS, UK).

(11) Land Instruments International, Wreakers Lane, Dronfield, Sheffield, S18 6DJ, UK.

(12) Temperature Handbook (Calex Instruments Limited, P O Box 2, Leighton Buzzard, Bedfordshire, LU7 8W2, UK).

(13) BS 368: Parts 4A to 4I *Measurement of liquid flow in open channels – weirs and flumes* (British Standards Institution, London, UK).

(14) BS 1042: Parts 1 and 2 *Measurement of fluid flow in closed conduits* (British Standards Institution, London, UK).

(15) ISO 5167-1: 1991 *Measurement of fluid flow by means of pressure differential device* (International Standards Organization, Geneva, Switzerland).

(16) ISO Technical Report TR 3313, 1992 (BS 1042: Part1: Section 1.6: 1993) *Measurement of pulsating fluid flow in a pipe by means of orifice plates, nozzles or Venturi tubes* (International Standards Organization, Geneva, Switzerland).

(17) BS 188 (1993) *Methods for the determination of the viscosity of liquids* (British Standards Institution, London, UK).

(18) **Gerrard, J. E., Staedtler, F. E.,** and **Appeldoorn, J. K.** (1965) Viscous heating in capillaries, the adiabatic case, *Ind. Engng Chemistry Fundamentals*, **4**, 332.

(19) **Polak, P.** (1971) Spontaneous hot zone formation in oil flow, *J. Mech. Engng Sci.*, **13**, 4.

(20) BS 89: 1990 *Direct acting indicating analogue electrical measuring instruments and their accessories* (British Standards Institution, London, UK).

(21) BS 90 (1993) *Specification for direct-acting electrical recording instruments and their accessories* (British Standards Institution, London, UK).

(22) **Bentley, J. P.** (1995) *Principles of Measurement Systems* (Longman Scientific and Technical).

(23) **Morris, A. S.** (1993) *Principles of Measurement and Instrumentation* (Prentice Hall Europe).

(24) **Higgins, R. A.** (1977) *The Properties of Engineering Materials* (Hodder and Stoughton).

(25) BS EN 10002: Parts 1 to 5 *Tensile testing of metallic materials* (British Standards Institution, London, UK).

(26) **Ashby, M. F.** and **Jones, D. R. H.** (1980) *Engineering Materials. An Introduction to their Properties and Applications* (Pergamon Press).

(27) BS EN 10045-1: 1990 *Charpy impact test on metallic materials – Test method (V- and U-notches)* (British Standards Institution, London, UK).

(28) **Stanton, T. E.** and **Batson, R. G. C.** (1920) On the characteristics of notched bar impact tests, *Proc. Inst. Civ. Engrs*, **211**, 91.

(29) **Griffith, A. A.** (1921) The phenomena of flow and rupture in solids, *Trans. R. Soc. A*, **221**, 163.

(30) **Wohler** (1871) (Various reports) *Engineering*, **11,** p. 349 and subsequent issues.

(31) **Taylor, T. E.** (1967) Effect of test pressure on fatigue performance on mild steel cylindrical pressure vessels containing nozzles, *Br. Welding J.*, **14**, 461.

(32) **Heywood, R. B.** (1962) *Designing Against Fatigue,* p. 432 (Chapman and Hall).

(33) Civil Aircraft Accident Report Comet G-ALYP, 10.1.54 and Comet G-ALYY 8.4.54 (1955) (H.M.S.O.).

(34) **Hammond , D. W.** and **Meguid, S. A.** (1990) Crack propagation in the presence of shot-peening residual stresses, *Engng Fracture Mechanics*, **37**, 2.

(35) ESDU Data Item 87007 (1987) *Design and material selection for rubbing bearings* (Engineering Sciences Data Unit, 251–259 Regent Street, London W1R 8ES, UK).

(36) BS 5514: Part 1: 1996 *Reciprocating internal combustion engines performance – standard reference conditions and test methods* (equivalent to ISO 3046-1: 1995) (British Standards Institution, London, UK).

(37) **Plint, M. A.** and **Martyr, A.** (1999) *Engine Testing Theory and Practice*, 2nd Edition (Butterworth Heinemann).

(38) **Dowling, D. M.** (1985) The evolution of grain edge porosity, Thesis, University of Surrey, Department of Physics, October 1985.

(39) **NAFEMS** (National Agency for Finite Element Methods and Standards) (1986) *A Finite Element Primer* (Department of Trade and Industry, National Engineering Laboratory, East Kilbride).

Appendix

Definitions of Instrument Terms

There are a number of definitions for the relationship of a transducer output to the measured value or 'measurand' given by the Instrument Society of America in its standard ISA-S37.1/1969. The standard claims only to cover transducers (devices which provide a usable output in response to a specified measurand) used in electrical and electronic measuring systems, although much of the terminology applies to other measuring instruments.

Accuracy is the descriptive term for the closeness of the measured value of a quantity to its actual value. It is better specified in terms of the maximum error.

Error is the algebraic difference between the indicated value and the true value of the measurand. It is usually expressed in percent of full-scale output, but sometimes in percent of the output reading.

Sensitivity is defined as the rate of change in transducer output to a change in value of the measurand, which is shown in Fig. A1 as the slope of the 'best straight line' through zero on the calibration curve.

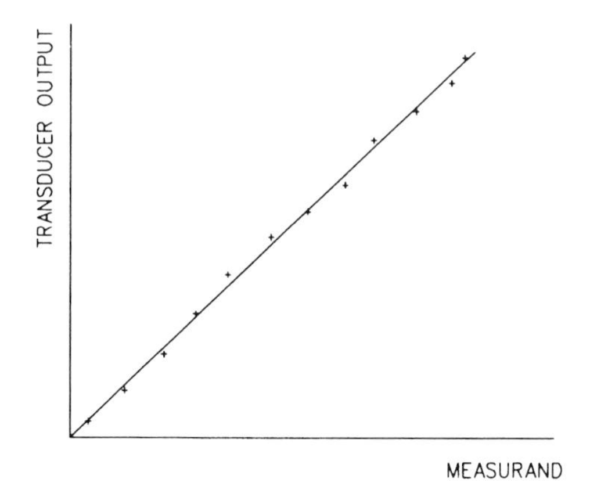

Fig. A1 Sensitivity; slope of the 'best straight line' through zero

Linearity is found by drawing two parallels (Fig. A2) as close together as possible but enclosing the entire calibration curve. Half the interval between the two parallels gives the linearity, which is less than or equal to +/– a percentage of the full-scale output.

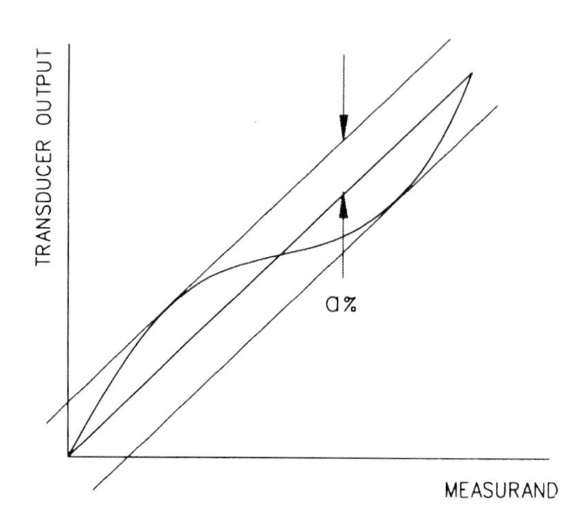

Fig. A2 Linearity; half the interval between the two parallels

Hysteresis (Fig. A3) is the maximum difference in the output signal for any value of the measurand within the specified range, when the value is approached first with increasing and then with decreasing measurand. Friction error is included in this unless 'dithering' (applying intermittent or oscillatory forces to minimize static friction) is specified.

Resolution of a transducer is the magnitude of output step changes as the measurand is continuously varied over the range (Fig. A4).

Threshold is the smallest change in the measurand that will result in a measurable change in transducer output. In one example, a piezo-electric force transducer, the resolution of the transducer itself was almost infinite, so the threshold was given as twice the value obtained at the charge amplifier by picking up noise on the connecting cable.

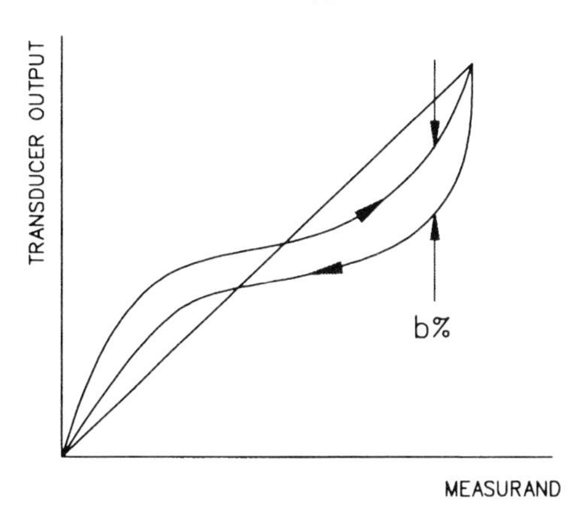

Fig. A3 Hysteresis; the greatest difference in output signal with first increasing and then decreasing measurand

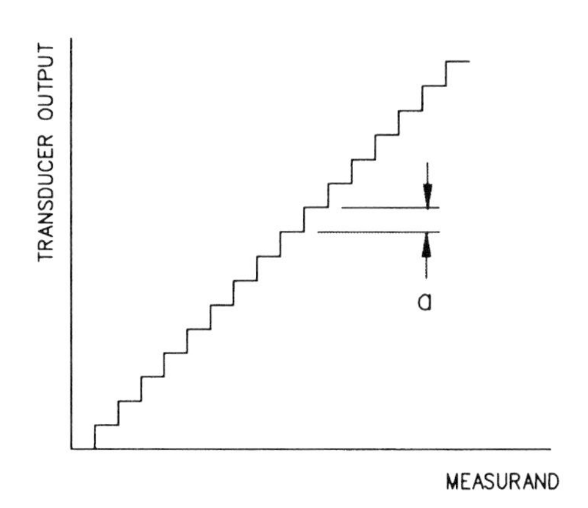

Fig. A4 Resolution; magnitude of the output step changes as the measurand is varied continuously

Zero shift is a change in the zero-measurand output over a specified period of time and at room conditions. This gives a parallel displacement of the entire calibration curve (Fig. A5). The zero can also shift due to changes in temperature or pressure, for example.

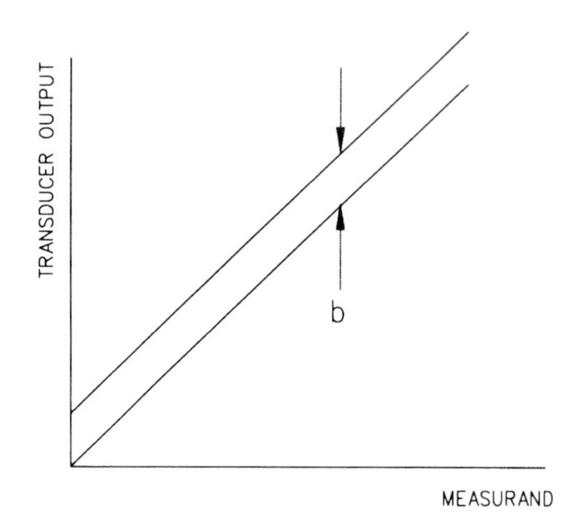

Fig. A5 Zero shift; a parallel displacement of the entire calibration curve

Stability is the ability of a transducer to retain its performance characteristics for a relatively long (specified) period of time.

Response time is the time taken by the output signal of a device to respond to a change in the input.

Noise in instrumentation is the term used to describe unwanted signals which appear in the transmission path and tend to obscure the transmitted measurement signal. Output noise is defined as the a.c. component of a transducer's d.c. output in the absence of measurand variations.

Scatter is the deviation from a mean value of precise readings.

INDEX